THE VEGETARIAN AGAMAS

AN UNCERTAIN FUTURE

In the terrarium hobby there are many species that are "old standbys," species that have been available year in and year out for decades. Other species always have been uncommon or virtually unavailable to the average hobbyist and continue to be just a dream for any hobbyist on a budget. Still other species come and go; one year they are rare and virtually unknown, the next they are everywhere. Occasionally such species become permanent fixtures in the terrarium hobby, but more commonly they disappear in a few years.

Presently we are seeing massive importations of a formerly very rarely seen agamid lizard, *Uromastyx ornatus*, commonly known as the Ornate Spiny-tail or Ornate Uromastyx. Adult males of this species are one of the most beautifully colored lizards, and the species is not hard to maintain. Unfortunately, this is the fourth species of its genus to be imported over the last several decades, and none has ever proved amenable to breeding in captivity. Presently, breeding of the Ornate Uromastyx is virtually impossible, but there is some hope that it will become more commonplace soon. The other three uromastyx species that have been imported on and off over the last decades still have not proved reliable breeders in captivity.

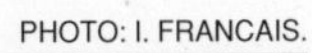

PHOTO: I. FRANCAIS.

A gravid female African Spiny-tail, *Uromastyx acanthinurus*. The attractive pastel colors are typical of warm, healthy specimens of these friendly, stocky lizards.

THE LEIOLEPIDINAE

This book is meant to be an introduction to the identification, care, and (what little is known of it) breeding of the African and Asian *Uromastyx* species and their close but rather different-looking tropical Asian relatives, the butterfly agamas, *Leiolepis*. These two genera share many features of their skull structure but have very different adaptations to living in dry, often desert-like surroundings. Together they are placed in the subfamily Leiolepidinae (formerly Uromastycinae) of the family Agamidae. They are closely related to the true agamas of the genus *Agama* and allies.

Both genera have short snouts on small heads that are covered with often tiny scales. The body is flattened to some extent, covered with small to tiny, granular scales, and at least the base of the tail is distinctly flattened. In neither genus is there any trace of a dorsal crest of erect scales as is found in many desert-adapted iguanas (such as *Dipsosaurus*). Butterfly agamas (*Leiolepis*) are very much like the North American whiptails (*Cnemidophorus*) in both appearance and behavior, having long tails and hind legs as well as a typically striped pattern. Spiny-tails (*Uromastyx*) are much like chuckwallas (*Sauromalus*) in shape and behavior, including the ability to change colors with body

BELOW: Adult spiny-tails eat an almost exclusively vegetarian diet. Try to provide a variety of leafy veggies, legumes, carrots, and even fruit-based commercial foods. This *Uromastyx acanthinurus* shows the reddish head often seen in the species. FACING PAGE: A couple of seven-month Ornate Spiny-tails, *Uromastyx ornatus*, chowing down on prepared iguana pellets. The bright colors of such foods may help stimulate feeding in these visually-oriented lizards. Photo: I. Francais.

PHOTO: I. FRANCAIS.

PHOTO: M. AND J. WALLS.

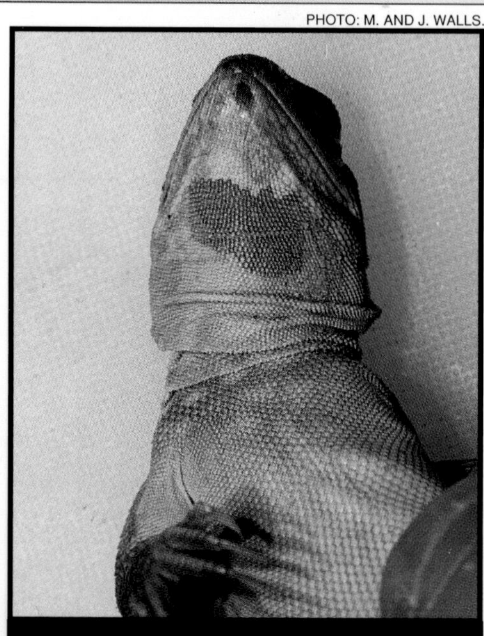

Underside of the head of a male Northern Butterfly Agama, *Leiolepis reevesi rubritaeniata*. The genera *Leiolepis* and *Uromastyx* together form the subfamily Leiolepidinae and appear to be closely related.

temperature, but they have whorls or rings of spiny scales on the tail more like the spiny iguanas (*Ctenosaura*). Both of these agamids are largely vegetarians, though both will take crickets, grasshoppers, and other insects when young and occasionally when adult. All are great burrowers that dig long, often complicated burrows that are difficult to accommodate under terrarium conditions.

CHECK THE TEETH

Like all the other agamids, the teeth of these lizards are acrodont, meaning they are solidly fused to the crests of the jawbones, rather than being in grooves on the inside of the jaws as in the iguanas (a condition known as pleurodont). Additionally, the teeth are fused at the base and look more like sawteeth than individual teeth. The uromastyx species have a uniquely developed set of teeth that are a strong adaptation to their vegetarian diet. Though hatchlings have four large incisor teeth at the ends of the jaws, with growth these teeth change considerably. First, the center two incisors of the upper jaws grow together to produce a single large median tooth directly at the center of the upper jaw, forming a type of beak. Often the incisor tooth behind and to each side of this median tooth fuses with it to increase the cutting edge. This sharp beak fits together well between the incisor teeth of the lower jaws to produce a shearing type of dentition that cuts through hard plant stems and leaves. Additionally, in the upper jaws the space behind the median tooth and incisors on each side forms a sharp, bony cutting edge in front of grinding, molar-like teeth that can dispatch hard grains such as rice and corn. Few reptiles have the teeth modified for different functions like those of the mammals, and the spiny-tails are one of the truly exceptional lizards in this regard.

INTRODUCING THE SPINY-TAILS

Because the thrust of this book is the spiny-tailed agamas or uromastyx, we'll leave the butterfly agamas for now and pick them up again at the end of the book. First, let's take a closer look at the uromastyx species and some of their basic requirements in captivity.

HOW MANY SPECIES?

Like many other lizard groups, no one knows exactly how many species of spiny-tails exist. The most recently published key recognized 14 species, but there has never been a complete revision of the genus and several species are so poorly known that their relationships cannot be determined with certainty. In the next chapter I'll briefly discuss each of the species and how to recognize them, so we can leave this survey for later.

THE TERRARIUM

All the uromastyx come from dry, desolate, often rocky plains that may be sparsely vegetated with tough shrubs and grasses (the steppes and savannas of the northern Middle East) or they manage to exist in nearly desert conditions, with shifting sands and little vegetation (deserts and dry plains of northern Africa and the Arabian Peninsula). In these circumstances they often are restricted to oases where vegetation is more abundant, but they never are dependent on water. In fact, under the nostrils of spiny-tails are large glands that help remove salt from the blood, often resulting in salt crystals accumulating around the nostrils on exceptionally hot days. In nature spiny-tails are active in hot midday temperatures, moving into their long and rather humid burrows about mid-afternoon and not coming out again until mid-morning.

PHOTO: M. AND J. WALLS.

From the side, the heads of spiny-tails and butterfly agamas are very similar. Notice the large vertically oriented ear opening, the large nostrils, and the prominent eyebrows. This is a male *Leiolepis reevesi rubritaeniata*, the Northern Butterfly Agama.

Because most uromastyx are large and bulky lizards that range from 10 inches to over 30 inches in length, you need a large terrarium at least 4 feet long and proportionately wide and high. These lizards do not climb much, but a secure mesh cover should be provided to help support the lighting system and keep out pests. A pair or small colony of one male and two or three females can be kept in a single terrarium if each is provided with its own burrow.The substrate may be sand, fine gravel, or even sod. Because it is impossible to provide a bottom layer deep

PHOTOS: I. FRANCAIS.

Two setups for spiny-tails. In the photo above, African Spiny-tails, *Uromastyx acanthinurus*, are provided with artificial burrows constructed of plastic boxes connected to flexible metal tubing, the entire burrow system being above the sand bottom. In the setup below, the smaller *Uromastyx ornatus* is housed in a simple sand-bottomed terrarium with hide boxes and no burrows. As they mature, they should be given access to burrows of some type.

PHOTO: I. FRANCAIS.

Detail of a simple burrow system for African Spiny-tails. Keeping the plastic box above the sand base makes it easier to inspect. The tubing serves as a burrow substitute and can be made of PVC, flexible metal, or even vent tubing as used for clothes dryers. Use whatever is convenient but be sure your animals are allowed access to tunnels and burrows.

enough to allow the lizards to dig their own burrows (you'd need a base at least 2 to 3 feet deep), most keepers provide sunken pipes of baked clay or PVC plastic as artificial burrows. Typically a burrow pipe should be about 3 inches in diameter (or a bit less with smaller species), at least 2 feet long, and have a 45 to 90° bend 6 inches or so from the end. The pipe is put into the substrate diagonally so the lizard can scurry down and build a resting chamber at the end. The top of the pipe should be flush with the surface of the substrate. Every lizard should have its own burrow to prevent fights, and it won't hurt to have an extra burrow in place for those rapid escapes that sometimes happen when any two lizards are confined together. All these burrows obviously dictate a terrarium with a large surface area and may be a factor

An under-tank heating pad under one corner of the terrarium helps maintain a constant temperature in part of the cage even during the night when the temperature is allowed to drop significantly. Photo courtesy of Fluker Farms.

restricting the number of animals you can house conveniently.

Some hobbyists have had improved success by planting a vertical pipe in each corner of the terrarium, perforating it with small drill holes in the bottom inch or two, and adding water once or twice a week. This simple standpipe helps keep the bottom of the terrarium moist while the top layer and air remain drier.

In still another method, a sealed plastic bowl is partially buried in the bottom and a hole is cut in its side at sand level. A damp paper towel (replaced twice a week or so) in the bowl provides the needed moisture.

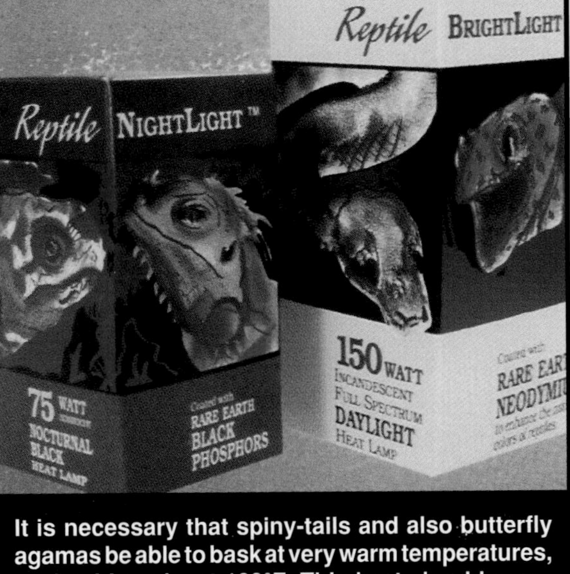

It is necessary that spiny-tails and also butterfly agamas be able to bask at very warm temperatures, preferably at least 100°F. This heat should come from above, not below. Hot basking lights should be used, often two or more in a terrarium, arranged over flat stones or some other basking surface. The bulbs will help the daytime air temperature stay comfortably toasty as well. Photo courtesy of Energy Savers.

LIGHT AND HEAT

The basic rule of uromastyx care is simple: BAKE 'EM. These lizards like an air temperature of at least 90°F from mid-morning to mid-afternoon and like to have at least one corner of the terrarium set up for basking at 100 to 115°F. At night the air temperature should drop drastically, preferably to 70 to 80°F, but the burrow must stay relatively warm.

Typically you'll need heat tapes or a heating pad under the terrarium, an incandescent basking light of some type in a reflector fitting, full-spectrum fluorescent lights over the terrarium lid, and perhaps a high-intensity light just to make sure that a warm enough temperature is attained for at least four to six hours a day. Provide fluorescent lighting for about eight to ten hours a day, preferably from perhaps 9 AM to 6 PM. Exposure to an ultraviolet light (UV-A plus UV-B) for ten minutes once a week seems to perk up the lizards considerably and bring out their best colors.

WATER

Under natural circumstances, it is unlikely that uromastyx drink water. They can get all the water they need by metabolizing their dry foods and also can get it from fat stored in the tail. Simply spraying their food with water should be more than sufficient to make the lizards happy. Some specimens like to be sprayed every week or so, but others dislike contact with water. (Under natural conditions in the desert, they may get some of their water from dew condensing on rocks and even their bodies.) A bowl of water may be used as a soaking bath by some spiny-tails, but this increases the humidity of the terrarium considerably and may be dangerous. There is no doubt that many African Spiny-tails, *Uromastyx acanthinurus*, like to soak and can absorb water like a sponge, apparently without consequences.

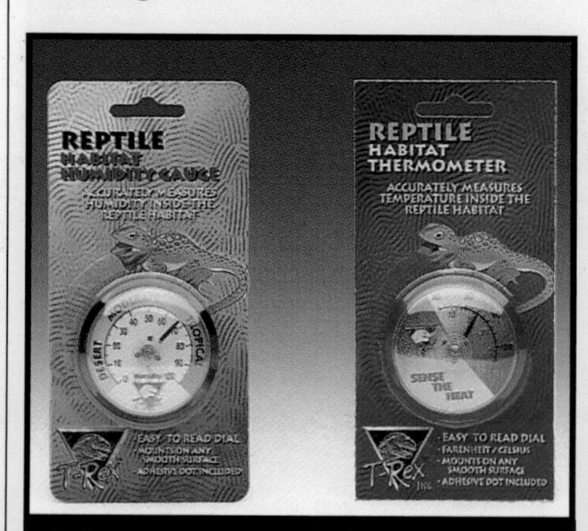

At least two thermometers should be in every terrarium at different levels and in the warmer and cooler corners. A small hygrometer (to measure humidity) will help alert you to when the moisture rises to dangerous levels (above 50% or so) in the terrarium. Photo courtesy of Ocean Nutrition.

Humid air is, on the other hand, dangerous to uromastyx. The high summer humidities of the northeastern United States and much of the southern U.S., for instance, may severely shorten the lives of all captive spiny-tails, even if they are given sufficient heating and fluorescent lighting. They may prosper for a couple of years, but then they stop eating and go into an irreversible decline. In very humid areas it may be necessary to play with dehumidified terraria during the summer in order to maintain your pets. Outdoor humidities over 50%, even if the temperature is in the 90's, may be deadly if continued over several weeks.

FOOD

Because spiny-tails are heavily parasitized when imported, be sure to have all specimens thoroughly vetted and wormed.

Though hatchlings and young uromastyx will chase and eat grasshoppers, beetles, and other insects, there is no doubt that these are vegetarian lizards. Adults do well on a diet of all types of leafy vegetables, including spinach, endive and other green (not iceberg) lettuces, cabbage (in moderation), broccoli, and clover. They also take green beans, peas, sweet corn (and water-soaked hard corn), rice (especially wild rice with the husk in place), carrots, apple peelings, and almost anything of a similar nature. They often like yellow and red flowers. Provide a daily feeding and make sure every lizard in the terrarium gets a chance to eat its fill.

Keepers are divided about giving animal protein to adult spiny-tails. An occasional treat of crickets, grasshoppers, mealworms, and perhaps even an earthworm or two probably does no harm and may increase the lizard's activity. However, a diet too heavy in animal protein may result in kidney damage (a literal "burn out" of the kidneys) and fluid retention that could lead to death. To be safe, give your pets insects only once a week at most.

With a little patience, many spiny-tails will learn to eat dry or semi-moist iguana food as either a stable diet or a supplement. These pellets provide a well-balanced diet and their bright colors may serve as a psychological incentive to feeding. Be sure to feed only fruit and veggie pellets, not animal protein. Photo courtesy of Ocean Nutrition.

By the way, these are among the tamest of lizards, and many specimens learn to feed from their keeper's fingers. Beware their strong claws and accidental bites, however; an enthusiastic lizard that weighs almost half a pound can inflict considerable damage with a near-sighted grab at food or a fast scuttle up your arm.

DETAILS

The general care scheme given above should work for most uromastyx. If your lizards stay well-fed and healthy, you have a good chance that they will breed or at least will lay eggs. There is good captive-care information on only four species of the genus, and we'll cover each of these in more detail in separate chapters to come. Don't expect too much from your lizards, however, just appreciate them for the large, somewhat clumsy, rather friendly, warm pets that they are.

A group of large metal stock watering tanks used as terraria for spiny-tails. Though these tanks can only be viewed from above, they are relatively inexpensive and deep enough to allow for a good substrate. Your pet shop should be able to get these for you if you are thinking of running a large number of spiny-tails.

PHOTO: I. FRANCAIS.

PHOTO: I. FRANCAIS.

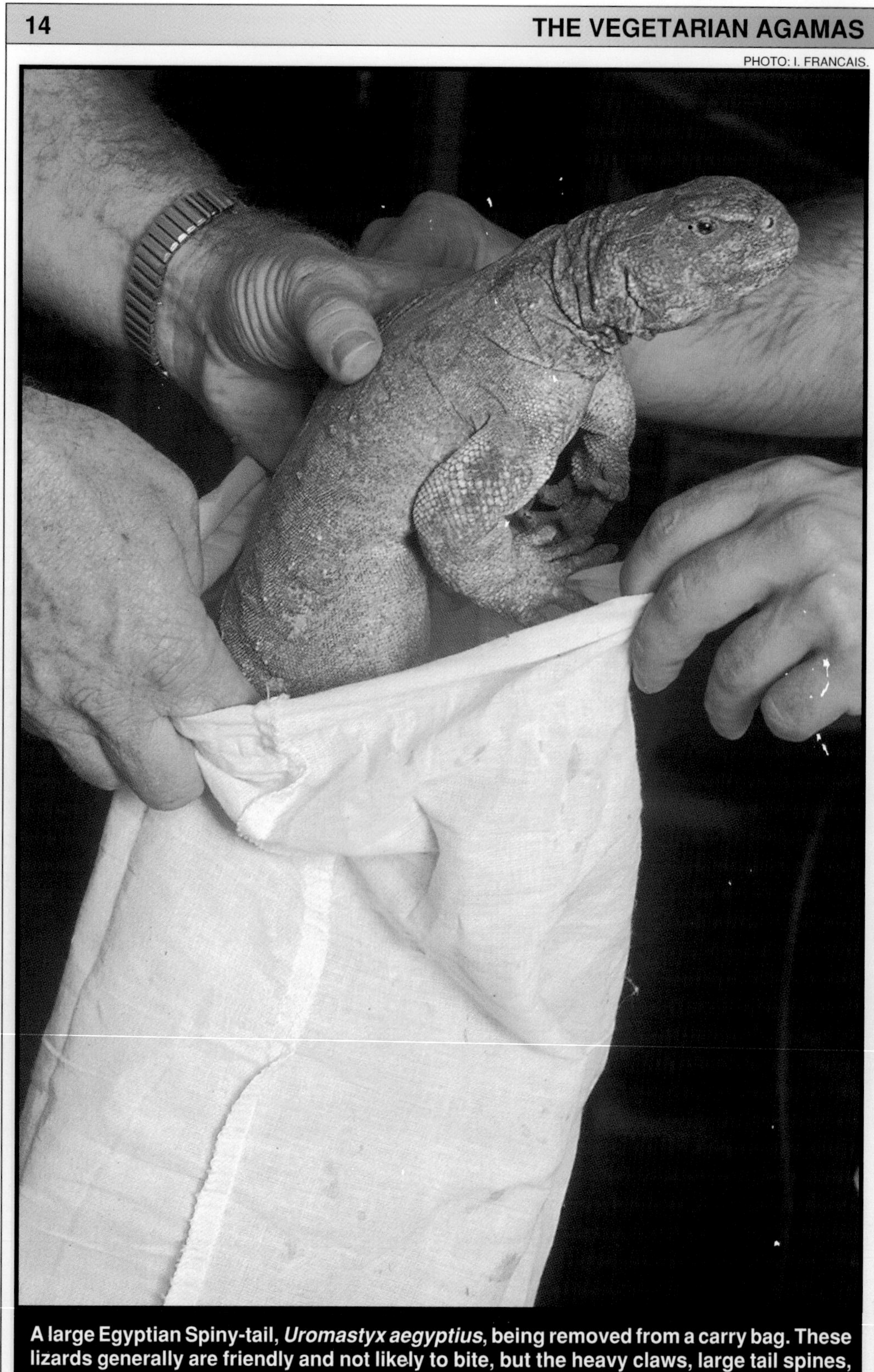

A large Egyptian Spiny-tail, *Uromastyx aegyptius*, being removed from a carry bag. These lizards generally are friendly and not likely to bite, but the heavy claws, large tail spines, and sharp thorns on the thighs may hurt if the lizard lashes at its handler or squirms and catches a finger between the body and leg.

PHOTO: I. FRANCAIS.

A convenient two-level basking surface for small *Uromastyx ornatus*: a patio block.

Tunnels and burrows mean safety to a spiny-tail, so they should be given some type of substitute. Use whatever is convenient for you, but make the lizards comfortable. Photo of *U. acanthinurus*:

PHOTO BY I. FRANCAIS.

SURVEYING THE SPINY-TAILS

Though almost 30 names have been applied to the species of *Uromastyx*, currently only some 14 are thought to apply to valid species. Surprisingly little is known of most of these species in either the wild or captivity, and several seem to never have been illustrated, even in the scientific literature. This is not really surprising when you consider that most of these lizards come from desolate, inhospitable terrain that is difficult to reach even today. Some of the species are restricted to areas that present problems of access because of politics (Iran, Iraq) or perpetual civil wars (Somalia, Yemen). In some ways it is surprising that even a few species of *Uromastyx* ever reach the terrarium hobby.

The genus usually is easy to recognize at even a glance, though it could be confused with some seldom-seen iguanas from South America that have similarly spiny tails, bulky and flattened bodies with fine scales, and fragmented head scales. None of these American lizards will agree in details of coloration and tail shape with any of the African and Asian *Uromastyx*, however, and a glance at the teeth should be the clincher, only *Uromastyx* having the large median tooth at the front center of the upper jaws.

Technically, *Uromastyx* has been defined as large lizards of the Old World that are depressed, lack a dorsal crest, have a gular (throat) fold but no gular pouch or fan, and have a distinct tympanal opening (eardrum). The nostrils are large and set on top of the short, blunt snout. The scales on the back are small, often granular, and typically smaller than those of the belly, while the head scales are small and seldom identifiable with the usual large scales present on the heads of most lizards. Salt glands are present and often produce salt crystals around the nostrils. The tail is short (usually less than half the total length), broad, depressed, and encircled with rows of large scales that bear high crests with free ends (spines). Except for two species, femoral and pre-anal pores are present in both sexes of all ages, each pore surrounded by a circle of tiny scales. (Obviously these lizards cannot be sexed by presence or absence of femoral pores as sometimes stated.) The modified teeth, discussed earlier, are the best character of the genus but of course may be hard to see in living specimens except for the median upper beak.

Synonyms of *Uromastyx* include *Mastigura, Centrocercus, Saara, Centrotrachelus*, and *Aporoscelis*. In addition to being called uromastyx and spiny-tailed agamas, they also have been known as dab or dahb lizards and mastigures.

The about 14 species have been placed in six groups, but there still is some controversy as to the real relationships of the various species, the results of studies based on structure and on biochemistry often giving different results. We have space here to just briefly discuss the distribution and range of each species, but admittedly except for the four species given separate chapters there is little natural history information available on the other species.

A CHECKLIST

Uromastyx Merrem, 1820

[Hardwicki Group]

U. hardwicki Gray, 1827 .... Indian Spiny-tail

[Asmussi Group]

U. asmussi (Strauch, 1863) .... Iranian Spiny-tail

U. loricatus (Blanford, 1874) .... Iraqi Spiny-tail

[Ornatus Group]

U. benti (Anderson, 1894) .... Poreless Spiny-tail

U. macfadyeni Parker, 1932 .... Somali Spiny-tail

U. ocellatus Lichtenstein, 1823 .... Smooth-eared Spiny-tail

U. ornatus Heyden, 1827 .... Ornate Spiny-tail

U. philbyi Parker, 1938 .... Yemeni Spiny-tail

[Princeps Group]

U. princeps O'Shaughnessy, 1880 .... Armored Spiny-tail

U. thomasi Parker, 1930 .... Omani Spiny-tail

[Aegyptius Group]

U. aegyptius (Forskal, 1775) .... Egyptian Spiny-tail

[Acanthinurus Group]

U. acanthinurus Bell, 1825 .... African Spiny-tail

U. dispar Heyden, 1827 .... Sudanese Spiny-tail

U. geyri Mueller, 1922 .... Saharan Spiny-tail

RECOGNITION

Because we'll discuss four species (the Indian, Ornate, Egyptian, and African) in some detail later, the following descriptions are meant to allow the other ten species to be recognized if they should be imported (*U. benti* and *U. ocellatus* already are imported occasionally).

RECOGNIZING THE GROUPS

If the tail of the uromastyx has small scales separating the whorls of enlarged spiny scales, then the lizard belongs to the Hardwicki (no large spiny tubercles on body) or Asmussi (large spiny tubercles present) Group. The remaining uromastyx have only large, very regular scales forming the whorls of the tail. The Princeps Group has the tail short (about half the head-body length) and broad, making them instantly recognizable. In the Ornatus

Group the scales under the posterior half of the tail are fused, not forming distinct whorls as they do on the top of the tail (as in the Acanthinurus and Aegyptius Groups). These latter two groups are very similar, but the Aegyptius Group has extremely tiny, granular scales on the back that are difficult to distinguish at a glance and almost impossible to count; in the Acanthinurus Group the scales are small but distinct. Lizards of the Aegyptius Group are almost uniformly brown from an early age (without distinct dark and light reticulations and barring in half-grown to adult specimens) and the form appears especially flattened, with many folds on the sides.

THE HARDWICKI GROUP

We'll discuss this large species from Indian, Pakistan, and Afghanistan more fully later, but for now it needs to be compared only with the Asmussi Group species, which also have small scales separating the spiny whorls on top of the tail. Unlike the Asmussi Group species, there are no enlarged spiny tubercles on the body. The tail spines are small and in many more whorls (34 to 36) than in any other species of the genus.

THE ASMUSSI GROUP

These two very unique species from the high plains of Iran, Iraq, and adjacent countries (Afghanistan, Pakistan) are similar to *U. hardwicki* in having the enlarged spine-bearing scales of the tail separated by rows of small, almost normal scales. *U. hardwicki* differs in having many more whorls of scales on the tail (34 to 36) than the Asmussi Group and a much less spiny tail.

U. asmussi from Iran, Afghanistan, and Pakistan has the scales at the front edge of the ear opening strongly spined or denticulated and has the loose skin of the nape and sides of the neck with many spiny tubercles. In addition, there are large, rounded, spiny tubercles in regular rows across the sides and extending onto the back. Often the tubercles are bright red in males, contrasting with the dark greenish brown back. This is a large species that reaches over 20 inches in total length. It digs burrows about 4 feet long and 2 feet deep, often with a distinct right-angle turn before the end. It is collected for both food (especially the fat in the tail) and its skin.

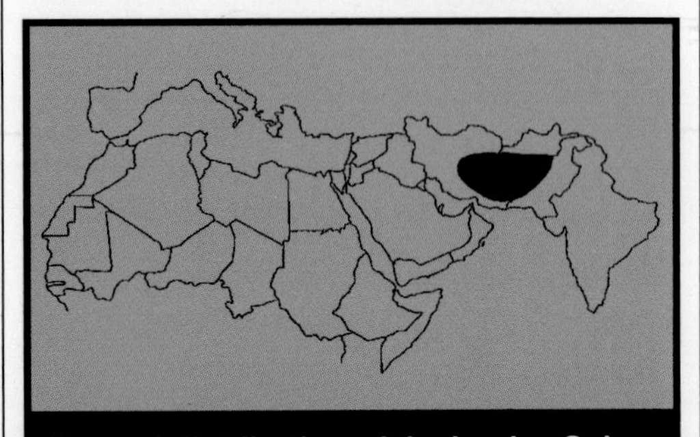

General distribution of the Iranian Spiny-tail, *Uromastyx asmussi.*

The very similar ***U. loricatus*** from Iran and Iraq also has spiny tubercles on the nape and spiny tubercles on the body, but these tubercles are not in regular rows and apparently are not red in males. The scales at the anterior

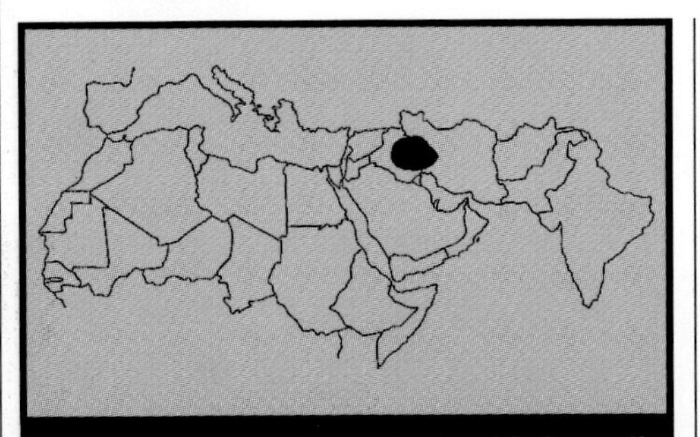

General distribution of the Iraqi Spiny-tail, *Uromastyx loricatus.*

edge of the ear opening are smooth, not denticulated, and the belly scales are somewhat smaller than those of *U. asmussi* (but hard to count). There are said to be 12 or 13 femoral and 3 or 4 pre-anal pores on each side, compared to 6 or 7 femoral and 2 or 3 pre-anal pores in *U. asmussi.* The coloration is paler, yellowish or cream above with brown spots and sometimes round yellow spots. There may be two brown stripes across the chest on an otherwise pale belly. They are active from at least mid-April to August in Iran and produce burrows up to 4 feet long in hard, rocky soil. Iraqi Spiny-tails are active from late morning to early afternoon during the hottest part of the day, and they have been collected actively feeding on the leaves of a local shrub at a soil temperature of almost 125°F. An active lizard may have a body temperature of up to 112°F, but they show distress by frequent panting with open mouths and rapid respiration when the body temperature rises above 113°F. Warm specimens become very pale, almost white with bright orange accents.

THE ORNATUS GROUP

Of the five poorly defined species of this group (the group is distinguished by having moderately large scales on the back, having the whorls of spines not separated from each other under the posterior half of the tail, and lacking projecting fringes on the outside of the fourth hind toe), one is recognizable instantly: ***U. benti*** lacks femoral and pre-

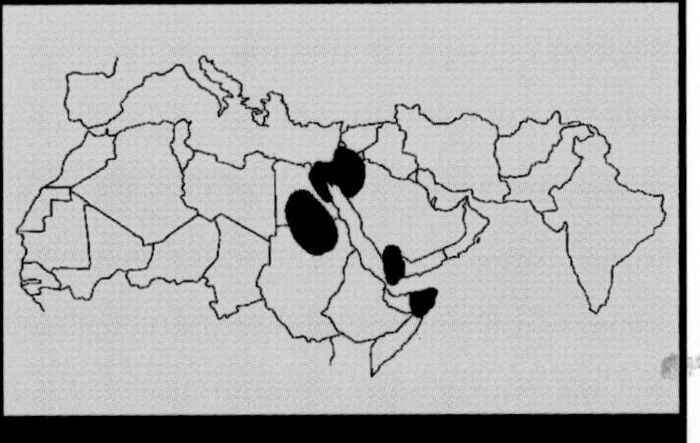

Overall distribution of the five species of the Ornatus group.

anal pores, a feature shared with only the very short-tailed *U. princeps.* The scales where the series would lie are roughened in males but totally lack pores. In pattern it looks much like *U. ornatus*, as do all the other species of this group. There are denticulated anterior ear scales, 23 to 27 whorls of scales on the

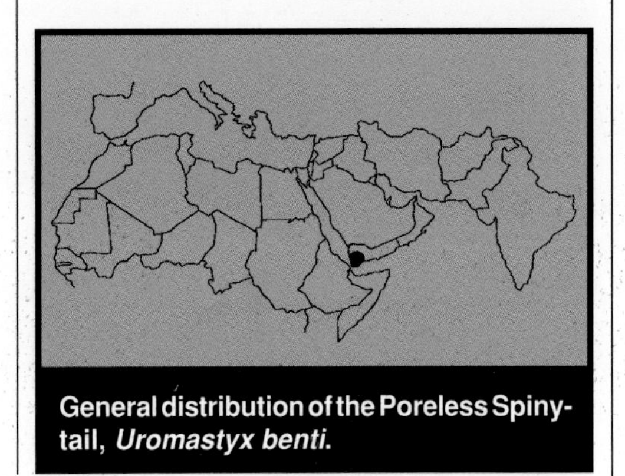

General distribution of the Poreless Spiny-tail, *Uromastyx benti.*

PHOTO: I. FRANCAIS.

An adult pair of Ornate Spiny-tails, *Uromastyx ornatus.* The male has assumed the bright blue-black coloration of his sex, while the female remains duller.

PHOTO: R. D. BARTLETT.

The Poreless Spiny-tail, *Uromastyx benti*, occasionally is imported from Yemen. Though superficially like an Ornate Spiny-tail, it lacks the femoral pores of its relatives. This is a male, as evidenced by the blue-black anterior sides.

tail, and the tail is over half the total length and strongly tapered. This is a species from southern Yemen that occasionally is imported but is virtually unknown in the literature. Adults reach at least 12 inches in total length.

U. ocellatus lacks denticulated scales at the front edge of the ear opening, present in the other species of the group. It is otherwise virtually identical in pattern and appearance to ***U. ornatus*** but comes from southern Egypt and the Sudan. There is only one row of scales in each whorl under the tail as in *U. ornatus*. These two species are very closely related and seem to replace each other geographically, and they often are considered subspecies (as *U. ocellatus ocellatus* and *U. ocellatus ornatus*). It reaches the usual foot or so in length.

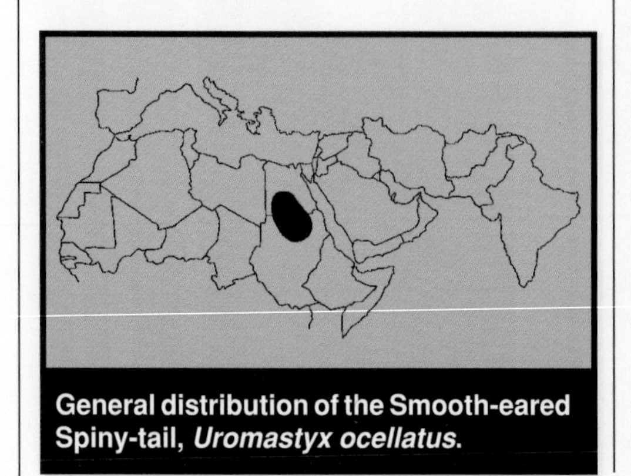

General distribution of the Smooth-eared Spiny-tail, *Uromastyx ocellatus*.

The Somali ***U. macfadyeni*** is again almost identical to *U. ornatus* but has two rows of

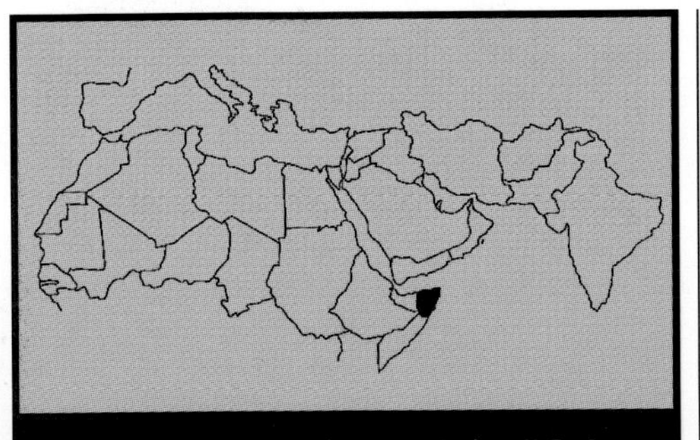
General distribution of the Somali Spiny-tail, *Uromastyx macfadyeni.*

scales in each whorl under the tail near the middle (one in *ornatus*) and has the spines at the side of the tail longer, almost equal to the diameter of the eye. There are about 9 femoral and 4 or 5 pre-anal pores on each side.

U. philbyi probably is a southern subspecies of *U. ornatus* (or of *U. ocellatus*, if you prefer) but has a broader tail that barely tapers through the first half (narrower and evenly tapered in *U. ornatus*). Adults are about 12 inches long and might have a more evenly reticulated color pattern than typical *U. ornatus*. It is known from the Jeddah area of northern Yemen.

It will be noticed that all five species of this group replace each other geographically as far as known and all could be subspecies of one wide-ranging species. The oldest name is *U. ocellatus*. As far as I can tell from the literature, none of the species can be told apart with certainty on the basis of coloration. The patterns vary considerably from individual to individual and locality to locality, perhaps responding to age, mood, and environmental and seasonal factors.

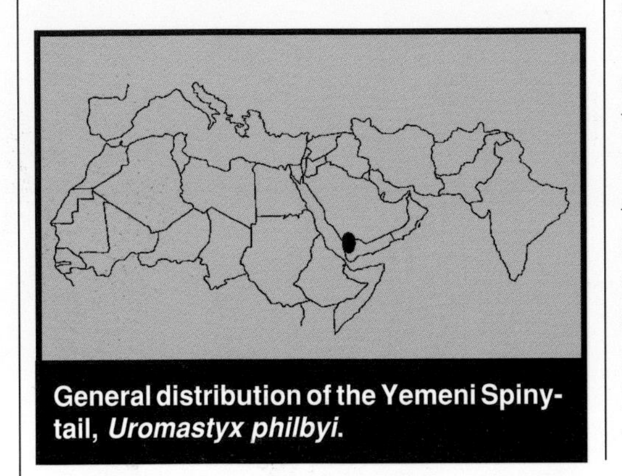
General distribution of the Yemeni Spiny-tail, *Uromastyx philbyi.*

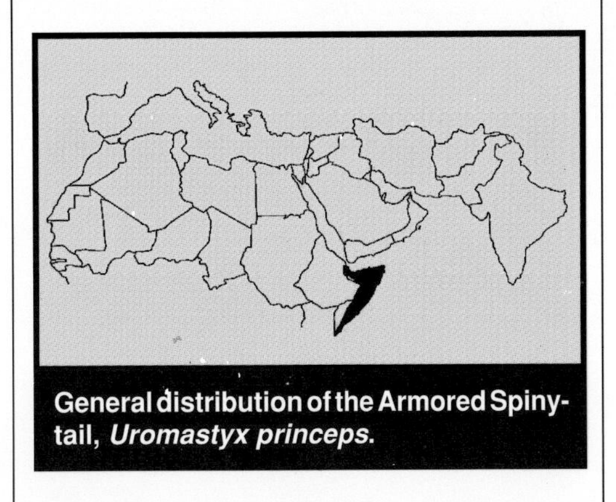
General distribution of the Armored Spiny-tail, *Uromastyx princeps.*

THE PRINCEPS GROUP

These are the short-tailed uromastyx, two species having the tail only about 50% of the head plus trunk length (SVL, snout-vent length) and rather disk-like. In ***U. princeps*** from Somalia and the Horn of eastern Africa the tail is a bit longer than in *U. thomasi* (50 to 60% of the SVL), the spines are extremely high and pointed, very bladelike, and there are no femoral and pre-anal pores. Adults are about 8 inches long and olive-brown above with small darker spots, becoming reddish on the hind legs and tail. *U. scortecci* Cherchi, 1954, from Eritrea is even more poorly known than typical Somali *U. princeps*

PHOTO: R. D. BARTLETT.

An immature Smooth-eared Spiny-tail, *Uromastyx ocellatus*. This species is very similar to the Ornate Spiny-tail, differing mostly in lacking denticulate (tooth-like) scales at the edge of the ear opening.

and may be a recognizable subspecies. There is an old record, probably incorrect, for this species from Zanzibar. The Armored Spiny-tail is found in areas with very rocky soil and appears to eat almost as much

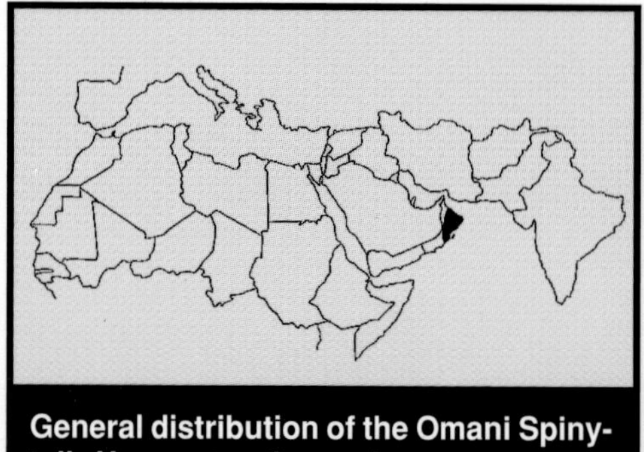

General distribution of the Omani Spiny-tail, *Uromastyx thomasi.*

insect food as vegetable matter in nature, but it is very poorly known and not likely to be imported in the near future.

The Arabian ***U. thomasi*** has the tail only 40% or so of SVL and distinctly disk-shaped with rather low spines. It has a total of at least 15 to 18 femoral plus pre-anal pores on each side, plus sometimes a short second row of pre-anal pores. Adults are only 6 to 8 inches long and may have a few dark cross-bars over the back plus large round black spots under the tail, or they may be almost uniformly tan reticulated with darker brown. It appears to be restricted to Oman, southern Arabia, including Masirah Island.

THE AEGYPTIUS GROUP

These large lizards have extremely tiny dorsal scales (over 300 along the midbody line) and have a spiny fringe on the side of the fourth hind toe. The single species, ***U. aegyptius***, is discussed in more detail later.

THE ACANTHINURUS GROUP

Often very colorful lizards marked with dark reticulations on bright yellow or green and with black heads and legs, this North African group of very similar species differs from the Ornatus Group in having the tail whorls distinct under the tail all the way to the tip. The scales in front of the ear opening are strongly denticulated. The lizards generally are adult at sizes over 12 inches, making them larger than typical Ornatus Group species. Until recently the group consisted of one species and about seven subspecies, but the last review (never fully published) recognized three full species with non-overlapping ranges. We'll discuss *U. acanthinurus* more fully later.

U. geyri from the Hoggar

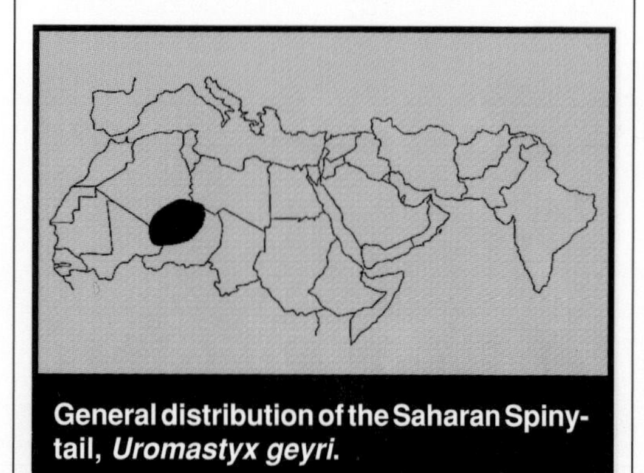

General distribution of the Saharan Spiny-tail, *Uromastyx geyri.*

Mountains of southern Algeria, Niger, and possibly Mali has 20 to 24 whorls of spines around the tail (only 17 to 20 in *U. acanthinurus*) and distinct

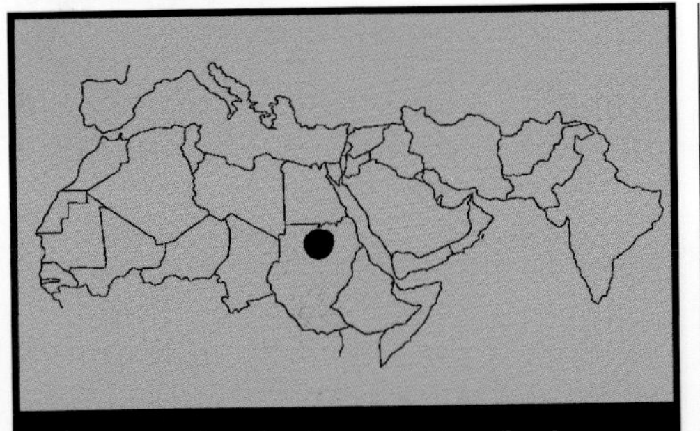
General distribution of the Sudanese Spiny-tail, *Uromastyx dispar*.

tuberculate scales on the vertical skin folds of the sides (only small scales like those of the back in typical *U. acanthinurus*).

The even more poorly defined ***U. dispar*** from the upper Sudan (perhaps still known from only five specimens collected at the beginning of the 19th century) has a brownish head with light spots, the back pale brown with dark spots, and the legs dark brown. The undersides of the body and tail have many distinct darker brown cross-bands against a paler background, supposed to be a distinctive feature of the species. (Many *U. acanthinurus* also have dark bands on a paler background, but they are more irregular and broken.) It perhaps averages smaller in size (14 inches total length versus often 16 inches or more in the other forms). Apparently most of the scale whorls around the tail have up to three rows of scales under a single dorsal row (typically two in *U. acanthinurus* and *U. geyri*). The possibility exists that this species is extinct, or that the few preserved specimens (most were reduced to skulls or dry skins) are not really typical of animals from the Sudan. Because of politics and economic problems, few collections from the Sudan reach the western world.

A colorful adult African Spiny-tail, *Uromastyx acanthinurus*. Notice the blackish legs, often a trademark of a warm specimen of the species.

PHOTO: I. FRANCAIS.

An adult Egyptian Spiny-tail, *Uromastyx aegyptius.* This generally dully colored species has finer scales than the African, a more oval head with a rounded snout, and more wrinkled, folded skin. Photo: I. Francais.

ORNATE SPINY-TAILS

For the last few years the American herpetological market has been inundated with a gorgeous uromastyx that previously was unknown in the mainstream hobby. If you look through the last 40 years of terrarium literature you won't find more than a casual mention of *Uromastyx ornatus*, the Ornate Spiny-tail, yet as of this writing it is even more commonly seen in the hobby than that old standby, the Egyptian Spiny-tail. Although dealers have plenty of young specimens and quite a few adults for sale, the Ornate remains a very poorly known species that cannot really be said to be established in the terrarium hobby. Many dealers will be glad to sell you captive-bred young, or so they will lead you to believe, but the reality is that almost no one has successfully bred the Ornate Uromastyx in the terrarium.

DESCRIPTION

One of the reasons for the popularity of the Ornate Spiny-tail is its small size. Unlike most of the other species we will discuss, adults seldom exceed 10 to 12 inches in total length, with the tail a bit less than half the total length. The head is short and rounded, with a large ear opening bounded at the front edge by enlarged scales that are distinctly toothed or denticulated rather than smooth. The scales of the head are small and irregular, and the skin on the neck is strongly folded or wrinkled. The scales of the back are small but obvious and easily counted, and those on the belly are a bit larger. The thighs bear enlarged, often rounded scales, some of which are spiny and may make the lizard hard to hold when it struggles. The tail is of course the crowning glory of this species, being wide, flattened, and heavily spined. If you look at the underside, you will notice that there is only one row of enlarged scales under each dorsal row, producing a complete simple ring of enlarged scales under the tail; the posterior-most ventral tail rings tend to be fused into a continuous armored club. There are about 12 femoral plus pre-anal pores on each side, the series almost running together across the body in both sexes. At least some mature males have larger femoral pores than females.

In coloration this species is quite variable individually, with age, and probably with geography. Juveniles are marked with wide dark brown bands across the back on a pale sandy tan background. The brown bands alternate with rows of fused yellow round spots that may be weakly ringed with darker brown. As the lizard grows these brown and yellow bands break down, resulting in the adult pattern. The adult pattern comprises about seven cross-bands of bright yellow spots and short bands outlined in a bright reddish brown. These yellow bands are very irregular and vary greatly from individual to individual, ranging from regular rows of circular spots to long bands that extend almost from side to side without breaks. Between the yellow spot bands are heavy reddish brown vermiculations (worm tracks, if you will) and dots in an extremely irregular pattern. There are several light (yellowish) and dark (black to greenish or blue) bars over the lower sides of the face, and the sides of adults are banded with alternating bluish green and yellow bars continuing the pattern from the back. Males may develop gaudy blue-green markings over the entire face, sides, and legs, much brighter than in females or immatures. The tail varies from plain brownish to heavily suffused with yellow and greenish blue. Ventrally, the throat and belly are covered with irregular brownish to bluish stripes extending from side to side but not covering the tail. In adult males the throat may be heavily striped with blue-black and pale blue. The head in adults may also assume a spectrum of colors in adult males, with a mixture of yellow, blue, and green scales. Males differ from females mostly by their brighter colors and sometimes wider heads.

Recognizing this species is not always easy. The five forms of the species group are virtually identical in general appearance, and color cannot be used as a definitive way of separating the species. So far only three species of the five are known to have been imported, and the other two species are rare and unlikely to show up. (See the survey chapter for characters of the virtually unknown *U. macfadyeni* and *U. philbyi*.) First check under the tail

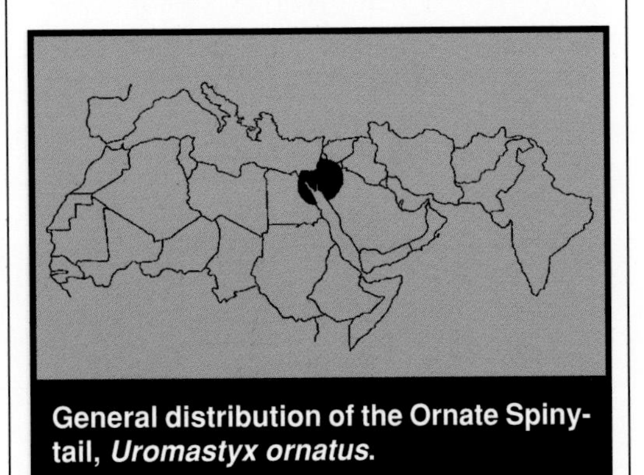

General distribution of the Ornate Spiny-tail, *Uromastyx ornatus*.

to make sure there is only one ring of enlarged scales under each whorl of spiny dorsal scales. This gets you to the right species group. If there are no femoral pores, then you have a specimen of *U. benti*, which occasionally is imported. If there are femoral pores but the enlarged scales at the front of the ear opening are smooth, without tooth-like tubercles, then you have *U. ocellatus*, which also is reported in the hobby. The five species of the group form a series of geographically non-overlapping forms that may represent only subspecies of a single species, *U. ocellatus*. The sequence is as follows: Somalia and adjacent areas of the Horn of Africa: *U. macfadyeni*; the Sudan and southern Egypt, *U. ocellatus*; eastern Egypt and Sinai, possibly to Syria and northwestern Arabia, *U. ornatus*; southwestern Arabian peninsula, Yemen, *U. philbyi*.

BASIC CARE

Its small adult size makes the Ornate Spiny-tail easier to care for than the larger species. They are friendly lizards that seldom fight if well-fed and allowed to each have their own burrow. Two or three adults can be housed in a 50- or 100-gallon all-glass terrarium with a strong lid to support the lights. Place a heating pad or heat tapes under one side of the cage where the burrows will end and put a hot basking light (100 watts or more, maybe even a metal halide lamp) at the other end so it shines over flat basking rocks. In between, bury several 3-inch in diameter PVC pipes so they extend from the surface of the sandy loam substrate to about 8 inches from the bottom of the cage, over or near the heating pad. If possible, put a right-angle bend in the pipe a few inches from the buried end. The lizards will adapt to these artificial burrows in a few days and build their resting chamber at the end. The chamber should be relatively moist compared to the very dry surface, so using a standpipe at the corners nearest the burrows and adding water on a regular basis works well. The air in the terrarium must be dry, and there is no need for a water bowl (an occasional soaking for a few

Detail of the head of a male Ornate Spiny-tail, *Uromastyx ornatus.* Notice the enlarged, tooth-like scales (denticulate scales) at the middle of the row to the front of the ear opening. This is almost the only character that separates it from *U. ocellatus.*

PHOTO: R. D. BARTLETT.

minutes, perhaps once a week, may aid in preventing shedding problems).

These lizards need lots of full-spectrum light, so use the proper fluorescent fixtures and change the bulbs often. You might want to use four lights rather than just two. A brief exposure (10 minutes or so) to a UV light once a week is said to increase color and perkiness.

This uromastyx is more agile than most, and there is a strong tendency among hobbyists to provide crickets and mealworms to adults in order to see a lively chase. However, these still are basically vegetarian lizards, and there is some concern that animal protein when fed to adults may result in kidney damage and shortened lifetimes. Feed your specimens a good variety of greens, root vegetables, grains, flowers, and the occasional pieces of fruit. All their necessary water should come from their food. The Ornate Spiny-tail will take some grains but is not as much a granivore as some of the larger species. Young specimens can be fed a mixed insect and veggies diet for their first six months or so, but then the animal food should be reduced and eventually stopped.

Like most of the uromastyx, it seems probable that the Ornate is long-lived in nature (at least 10 to 12 years in the large species), but so far few specimens have really matured in captivity. Immatures, especially, will die quickly if kept too cool, so make sure they can bask at about 95 to 100°F (and

PHOTO: W. P. MARA.

Underside of a mature male Ornate Spiny-tail, *Uromastyx ornatus*. In this specimen the femoral pores are enlarged and brown because they are secreting a wax-like substance that the male uses to mark his territory. The color pattern of the belly varies greatly in this species and other spiny-tails.

hotter is better), with a general air temperature of at least 80°F. Let the cage temperature drop to 70 to 75°F at night, following the normal desert rhythm of hot days and cool nights. These animals react badly to long periods of warm, humid, stuffy air, making them difficult to keep successfully during the humid summer months in much of the United

States. You may have to attach a dehumidifier to their cage, run an air conditioner and increase the heat in the cage, or perhaps use desiccants such as silica gel in the air flow into the terrarium.

BREEDING

So far relatively little definite can be reported about breeding *U. ornatus* in captivity because there have been few successful attempts by hobbyists as of this writing (though a few zoos have succeeded on a limited basis). It can be assumed that the species breeds like the other members of the genus. Following a brumation of about two to three months at 65°F, the specimens will come out ready to mate. Only healthy individuals with much fat stored in the tail should be allowed to brumate, and even they should be given a small basking light for a few hours each day in case they become temporarily active. Males will court females by whirling in place and marking a territory, followed by biting at the nape and anterior sides. Eggs should be laid about a month after mating and normally would be held in a chamber off the burrow. They may prove difficult to incubate, like other uromastyx eggs.

PHOTO: W. P. MARA.

A colorful juvenile Ornate Spiny-tail, *Uromastyx ornatus*. This is the size and color pattern usually offered for sale.

THE FUTURE

Perhaps by the time you read this there will have been a breakthrough in breeding this species, and truly captive-bred young will be readily available for a price. At the moment, however, this seems unlikely, and the hobby will continue for at least a while to be dependent on wild-caught imports reportedly from Egypt. There the lizards are both collected from the wild and "ranched," gravid females being held until they lay in natural burrows and the eggs hatch, with the young being shipped to markets in America, Europe, and Japan. However, recently various types of political and economic pressures have been brought against Egypt to force her to reduce or stop exports of some herps that are being exploited in tremendous numbers. The Sudan and Israel export few herps, even if political conditions allow such a trivial activity to continue. If this species is not bred soon in good numbers, it may not be around at the turn of the century. The same can be said about its close relatives that

PHOTO: I. FRANCAIS.

Above, a pair of adult Ornate Spiny-tails, *Uromastyx ornatus.* The blue-black sides and face of the male, combined with the bright colors of the back, make this one of the more attractive Old World lizards. Below, A portrait of a male Ornate Spiny-tail, *Uromastyx ornatus*, that is not quite in full color. Notice the pale blue scales on the face and the complexity of the dorsal pattern. On the tail, notice that there are no small scales between the rings of spines.

PHOTO: R. D. BARTLETT.

PHOTO: I. FRANCAIS.

Portrait of a male Ornate Spiny-tail, *Uromastyx ornatus.* In this specimen the head is almost all blue-black and the denticulate ear scales are easily seen.

are imported even less frequently and are unknown to most hobbyists. This is a great species to form the basis of an experimental breeding program for the fairly advanced hobbyist. The animals are relatively inexpensive for such a brilliantly colored lizard and don't take that much room to house compared to larger species. They certainly are worth a try—but do it soon before it's too late.

Two juvenile (seven months old) Ornate Spiny-tails, *Uromastyx ornatus.* Notice the great difference in the back coloration and pattern. Details of coloration are almost meaningless in this genus of lizards.

PHOTO: I. FRANCAIS.

AFRICAN SPINY-TAILS

Uromastyx acanthinurus is ***the*** spiny-tail of the German terrarium literature, specimens having been exported from Morocco and Algeria to the European market for decades. It is a big, personable lizard that is easy to keep but difficult or virtually impossible to breed with regularity under typical hobby conditions. A species of the Sahara and its mountains and oases, the species ranges widely from northern Senegal to Algeria, with two very closely related species just to the south and east of this basic range. In the northern part of its range it may be abundant, and until recently it has been a staple export from the Moroccan herp markets. At the southern edges of the range, however, it may be very rare, with only one or a few records of specimens from isolated oases where there is sufficient moisture and plant life to allow life to continue. In the last few years new environmental laws in Morocco and political unrest in other North African countries have reduced or eliminated fresh exportations to the terrarium market.

DESCRIPTION

The African Spiny-tail is large, adult males often exceeding 16 inches in total length, with the heavy, spiny tail comprising a bit less than half of the total length. Females are somewhat smaller, at least in most populations, and tend to have narrower heads than males. The head is short and blunt as usual for the genus, the nostrils are large, the scales are irregular, and the enlarged scales at the front edge of the ear

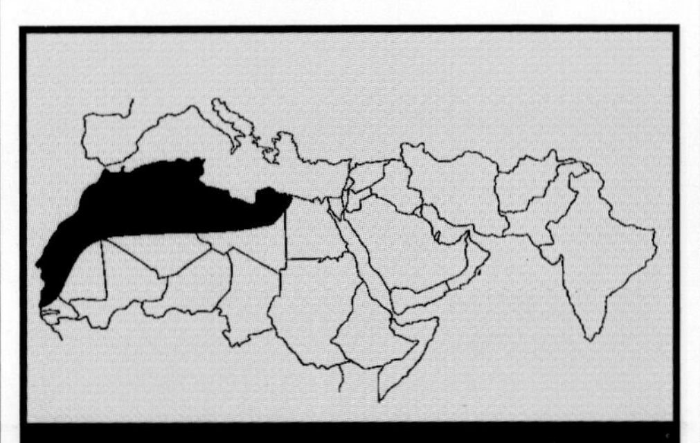

General distribution of the African Spiny-tail, *Uromastyx acanthinurus.*

opening are denticulated. The scales of the back are small, but still countable without much difficulty, and the sides have obvious wrinkles or folds of skin, especially prominent behind the head. The legs are large and strong, especially the hind legs, with heavy claws. There are heavy, spiny scales on the thighs. The tail is heavy, often with nearly parallel sides, and quite spiny, but the spines are relatively short compared to some other species of the genus. Typically there are about two rows of scales under each enlarged dorsal spiny scale row or whorl, but there appears to be quite a bit of variation in this ratio.

The coloration of *U. acanthinurus* is difficult to summarize because there is a great deal of variation, some of which may be geographic in

PHOTO: I. FRANCAIS.

A gravid female African Spiny-tail, *Uromastyx acanthinurus.* Color variation in this species is enormous, differing with age, mood, locality, and individual.

origin. Until recently, six subspecies were recognized in this species. Today two of these have been elevated to species rank (*U. geyri, U. dispar*) though they seem marginally distinct. The other names have been placed in the synonymy of *U. acanthinurus*, which now is without subspecies, though the details of the study leading to this move have not yet been published.

Typically, this is a colorful lizard. The head and legs tend to be dark brown to almost black, while the back is grayish, reddish brown, or olive, heavily covered with a network of black or dark brown. Often the network is very irregular (worm tracks or vermiculations), but just as

A familiar color pattern in the African Spiny-tail, *Uromastyx acanthinurus*, consists of a pale (often yellow or greenish) body and black legs and tail.

PHOTO: J. COBORN.

commonly it is more regular, forming large rounded meshes that give the effect of large bright olive spots. The tail is brownish, often marked with reddish or black spots. When warmed up to full activity or exposed to UV light, the colors become considerably brighter, and the back may turn bright yellow or green. Commonly there are a few dark stripes down from the eye over the jaws. The belly is typically whitish or pale yellow, with or without irregular narrow brown bands or meshwork. Females may show a tendency to have cleaner, whiter bellies than males, but it also is not uncommon for the undersurfaces of the body to be almost black. Many males when ready to breed show reddish heads. Hatchlings (about 3 inches long) are pale reddish brown above and white on the sides and belly, with few darker markings.

Because the dark meshwork on the back may be so variable in development, many different patterns are possible. One of the most extreme is the development of wide broken black and yellow bars across the back. Other specimens have many perfectly developed large yellow spots ringed in blackish brown, and still others are almost uniformly black over the back as well as the legs and head. This variation led to the description to several subspecies. As mentioned, *geyri* and *dispar* now are considered distinct species and the other names are synonyms. For the record, the previously recognized subspecies were: *U. acanthinurus acanthinurus* Bell, 1825: northern Algeria and Tunisia, questionably eastward into Egypt; *U. a. flavifasciatus* Mertens, 1962: northern Senegal into Mauritania and southern Morocco; *U. a. nigerrimus* Hartert, 1913: southern Algeria; *U. a. werneri* Mueller, 1922: Morocco and adjacent Algeria. Most of the specimens in the terrarium came from Morocco and Algeria.

BASIC CARE

Big lizards need large terraria, especially when you keep two or three specimens together. Try to give them at least 4 feet of ground surface and a substrate 2 feet deep. These lizards are at home both in the usual sandy loam and in sand, but they cannot burrow satisfactorily in loose sand. If you provide them with many rock piles on the surface they may use these as well as the burrows. Give the lizards PVC pipes of appropriate diameter and length with a sharp bend above the end if you wish to reduce the disturbance caused by the lizards constantly burrowing. African Spiny-tails often are found near oases, so it is best if their burrows are somewhat moist; use a standpipe filled with water occasionally to keep the bottom layer of substrate moist. The top of the substrate and the air in the terrarium must be kept dry, however, as these lizards cannot thrive in moist air.

As with all the other uromastyx, provide the cage with a bank of at least two (preferably four) full-spectrum fluorescent lights made specifically for reptiles and a 100-watt or better basking light in a reflector. You must keep the air temperature during the day at 85°F or more, with a basking area at 100 to 120°F for at least eight hours a day. At night, turn off the lights and let the terrarium drop to room temperature. If you want to keep these big lizards outside, make sure they can get into shade when necessary and do not leave them out when the humidity rises to more than 70%. They are strong and constant burrowers and will go through flimsy wood and mesh walls of a temporary outdoor cage.

Adults are largely vegetarian, taking a great variety of vegetables, flowers, and grains. Variety is the name of the game to provide a balanced diet. As usual, immature specimens will take insects of all types, but adults should be fed a minimum of animal protein to prevent possible kidney problems. Water is not necessary, but many specimens enjoy a short soak every week or two and will absorb water like a sponge. Remember the problem with high humidity, however, before deciding to soak specimens on a regular basis. Some keepers suggest that this species needs a great amount of mineral supplementation, not just calcium but also various earth salts. It might be best to provide this species with regular vitamin and mineral supplementation each week.

BREEDING

The Germans have kept hundreds of African Spiny-tails for decades, the specimens often surviving for many years but seldom breeding in captivity. Even zoos, which have relatively unlimited funds and manpower to devote to breeding odd animals, have had only sporadic successes breeding *U. acanthinurus*. First, start with a brumation period of two to three months at 65°F. During this time the lizards may be relatively active, and you should provide them with a small basking light and small meals when they come up to take the air. When the adults come fully out of brumation in February or March, they should be ready to breed. The male marks a territory by curling into a tight circle, the tip of the tail being held next to the snout. He then spins in place and also deposits a white substance from his cloacal glands on the ground. This display may continue over several days if an interested female is present. It is not uncommon for females and even immatures to also spin in place, by the way. If a female is ready to mate, she allows the male to bite at her neck and sides without running away, finally allowing the male to hold her in position to insert a hemipenis into her cloaca.

Egg-laying occurs about four to six weeks after mating. In nature the eggs are laid in a chamber off the burrow or under a rock on the surface. The soil in which the eggs are laid may be only 85°F, so

PHOTOS: I. FRANCAIS.

obviously they should not be laid in the warmest part of the terrarium. A clutch may consist of a dozen to almost 20 eggs over an inch long, white, and oval in shape. They can be incubated in moist vermiculite or sand in a relatively humid incubator. A temperature of about 86°F may (or may not) produce successful hatching in about 100 days. The young often are very weak at hatching, may not feed for several days, and have a miserable survival record. I think it is safe to say that very few hobbyists have ever been able to breed this species and successfully produce young year after year.

The high eyebrows and large nostrils are typical of several spiny-tails. In the African, *Uromastyx acanthinurus*, the snout is relatively narrow and almost pointed. The color changes rapidly with temperature and mood, with warm, active specimens sometimes being extremely colorful.

An exceptionally colorful young African Spiny-tail, *Uromastyx acanthinurus*, with a pattern very similar to that of the Ornate Spiny-tail. The head is a bit wider and the snout more pointed than in that species, the scales of the back are finer, the tail is a bit more depressed down the center, and the coloration is greenish, unusual in the Ornate. Photo: A. Norman.

PHOTO: I. FRANCAIS.

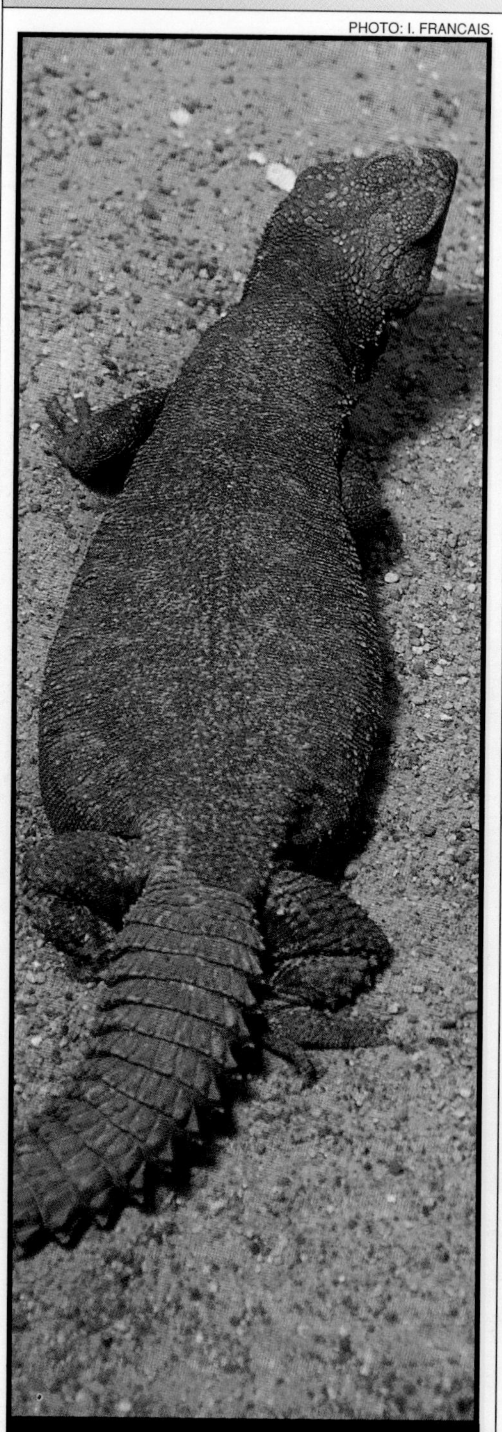

Not all African Spiny-tails are brightly colored. This *Uromastyx acanthinurus* is as dark as an Egyptian Spiny-tail, but notice the more pointed head, less depressed body, and lack of strong wrinkles on the sides and back.

THE FUTURE

At the time of writing, most importations of *Uromastyx acanthinurus* have disappeared. Morocco, the main source a few years ago, and Algeria both have restricted exports of their fauna, and there are no signs that the laws will loosen up in the near future. (Of course, as with all things political, tomorrow hundreds of African Spiny-tails might be exported.) Continuing droughts and economic and political problems throughout northern Africa have devastated some of these countries, and it is likely that marginal populations of African Spiny-tails may now be extinct because of drought and predation by hungry locals. Like other large lizards, these uromastyx are considered quite edible in times of hardship, and the skins also are salable. Few breeders have been able to maintain a line in captivity or produce excess young for sale, and zoos apparently do not have a reservoir of specimens that might eventually reach the hobby market. Fortunately, one or two breeders in the southwestern United States have succeeded in producing fair numbers of captive-breds, but they are expensive and far from easy to find. If you should find any African Spiny-tails for sale you might be wise to purchase them, have them carefully vetted and treated for worms, and give them the best conditions you can. They could be breeding stock for the last of these lizards to become available to hobbyists.

EGYPTIAN SPINY-TAILS

This large, brown, quite flat spiny-tail is perhaps the most readily available species at the moment in the United States (subject to change without notice, of course). Large numbers of adults and immatures are being imported to sell at rather low prices. Unfortunately, their condition seldom is acceptable, and few pet shops seem to be able to maintain the species at its required temperatures. Probably few of the imports survive for long, but this does not have to be.

At 24 to 30 inches in length and robust in body build, this is a large lizard with weights approaching 2 pounds. It occupies dry, often salty plains and deserts from Egypt to southern Syria and east to Iran along the Persian Gulf, including most or all of the Arabian Peninsula. There it burrows in loose soils, often near oases, and feeds on dry vegetation and seeds. The very heavily spined tail and enlarged spiny tubercles on the thighs protect it from most predators, though like other large lizards it is hunted and eaten locally and also taken for its skin.

DESCRIPTION

The Egyptian Spiny-tail is easily recognizable by its very fine scalation. There are well over 300 scales down the midline of the back, more than in any other species. However, you don't have to count scales because their granular nature makes the back appear almost scaleless. The body of this species appears to be especially depressed, and the skin of the sides especially wrinkled, the folds often extending well onto the back. The head is small, blunt, and covered with small and irregular scales that are largest on the snout and over the eyes. The ear opening is large and has a row of weakly denticulate scales at its anterior edge. The scales of the back are finely keeled; those of the belly are smoother and only slightly larger than those of the back. The hind legs are very heavy in adults and have a row of large, spiny tubercles on the thigh. The tail tapers but is broad and heavy at the base, with large spines above and two or three rows of scales under each dorsal row; it is shorter than the head and body length, as usual. There are up to 20 total femoral and pre-anal pores.

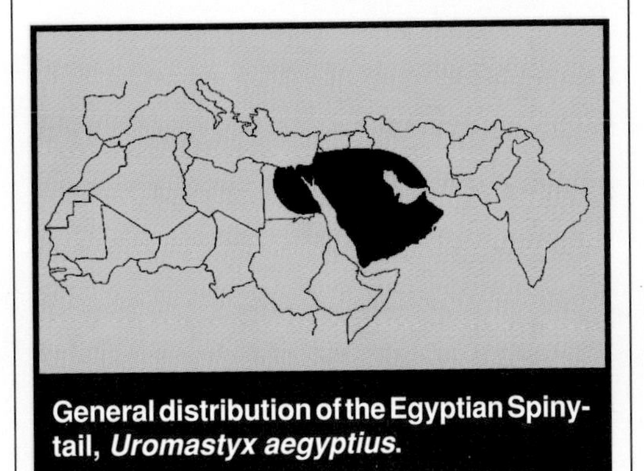

General distribution of the Egyptian Spiny-tail, *Uromastyx aegyptius*.

As used here, the Egyptian Spiny-tail, *Uromastyx aegyptius*, may include two distinct species. Often the species is broken into two taxa on the basis of the presence or absence of spiny tubercles on the skin folds of the sides. Specimens from Egypt, the Sudan, the Sinai, and perhaps further east (to Iraq?) have spiny tubercles mixed among the regular scales of the sides. These are the typical *U. aegyptius*, once known by the synonym *U. spinipes*. Over much of the Arabian Peninsula, north to Iraq and Jordan and east to western Iran (on the Persian Gulf), the lizards have slightly smaller scales on the back and lack spiny tubercles on the sides. This form long has been known as *U. microlepis* Blanford, 1874, but there has never been a large-scale published study of variation in this group of lizards. Presently some authors consider *U. microlepis* to be a subspecies of *U. aegyptius*, while others consider the presence or absence of spiny lateral tubercles to be an individual variation that is not completely related to distribution, making *U. microlepis* a synonym of *U. aegyptius*. Until details of variation in this species are published, its taxonomy will remain in dispute. The

Typical adult Egyptian Spiny-tails, *Uromastyx aegyptius*, are dull gray or blackish with many strong wrinkles running from the sides onto the back and a blunt, rounded snout.

PHOTO: R. G. SPRACKLAND.

specimens in the hobby generally seem to be coming from the western part of the range and usually have small spiny tubercles on the sides as would be expected.

In color this is not an especially exciting species. Most adults are dark brown on the back with darker legs and head. The tail is dark brown like the body, the belly paler and often heavily marked with a brown meshwork. However, if specimens are allowed to warm up to natural temperatures they become a bit more colorful, often gaining an olive tinge on the back and bluish green mottling on the side of the head. Such warm specimens may be more heavily spotted with dark brown on the back. Two types of juvenile coloration seem to be present in imported specimens (perhaps related to geography?). Most young are simply dark brown with a few rather irregular small yellowish or white spots on the back. Other specimens (including juveniles from Arabia recorded in the literature and corresponding to *U. microlepis*) have a more elaborate pattern resembling a reduced *U. ornatus* pattern. There are six to eight narrow yellow bands extending across the back, each band broken into a mixture of large oval spots or short stripes, usually with a large rounded spot at the center of the back in each row. This pattern is present in specimens up to at least 7 inches in total length.

PHOTO: I. FRANCAIS.

Belly view of a large male Egyptian Spiny-tail, *Uromastyx aegyptius.* The small head with very large nostrils is typical of the species. The femoral pores are small and rather indistinct except in breeding males.

BASIC CARE

Like the other species of spiny-tails, you have to bake this species to make it happy. Because of its size it needs a large terrarium deep enough to accommodate a substrate depth of at least 2 feet. This species is not always as complacent and gentle as the others of the genus, and large adults may fight over food, but two or three adults usually can be kept in a large cage. Several zoos have noted that adults take at least two weeks to acclimate to their cages before beginning to feed.

Because almost all specimens on the market are wild-caught, be sure to take your new pet to a vet and have it checked and wormed—adults may have large numbers of parasites. Because many specimens have been kept in substandard conditions at low temperatures, be especially on the lookout for respiratory infections. Often soaking a newly obtained specimen for a few hours in lukewarm water will perk it up and also give it a less wrinkled body shape. These are very hardy lizards, and if you immediately give your new acquisition proper heating, lighting, and food it may survive to a ripe old age.

Given the opportunity, Egyptian Spiny-tails will burrow like any other uromastyx. Because of their large size and wide body, they may need larger PVC tubes than for most other uromastyx, so match a 2- or 3-foot length of PVC to the body size of your pet. Give the pipe the usual sharp turn above its end and bury it in the substrate so the upper end is nearly flush and the buried end stops above the bottom of the cage with enough room for the lizard to make a living chamber. Also provide a hidebox or two and several other hiding places at the surface, along with flat rocks under the basking light. These lizards have been more-or-less successfully kept outside in metal cattle watering tubs and similar containers half-filled with a mixture of soil, gravel, and wood chips. Beware of summertime humidity, however, in animals caged outdoors.

Heating and lights should be as for other uromastyx, an air temperature of about 85°F or more, with a basking temperature under a hot incandescent or metal halide lamp of at least 100°F and preferably 110°F or more. The burrow should be cooler and more moist than the rest of the terrarium. Let the temperature drop each night to room temperature. A short exposure to UV light each week may perk up the animals. The air in the cage of this species should be as dry as possible.

Food for adults should be the usual vegetable salad with as broad a variety of greens and veggies as possible. Try almost anything, from leafy veggies and dandelions to grains such as lentils and wild rice. Don't forget the flowers, bean sprouts, and clover. Some specimens have done very well on a good grade of hay. Water may be given as a quick soak once a week, but no water bowl is necessary as the food provides all the lizard's water needs. Immatures may take crickets, grasshoppers, mealworms, and waxworms mixed with their vegetables, but this high-protein diet generally is not considered to be suitable for adults.

BREEDING

Success in breeding the Egyptian Spiny-tail has been limited, as with all uromastyx. Most breedings have followed a brumation (limited hibernation) period of a few months at about 65°F. Normally the specimens are gradually subjected to shorter day lengths and lower temperatures through November and the lizards are brought out of brumation in February. Like other uromastyx, their hibernation is not complete, and they often come out of their burrows for an hour or two on warm days through the winter. A small basking light may help them warm up a bit on occasion, but never let the temperature become as warm as during the active part of the year.

Mating should occur when males and females are put together after coming out of brumation. Males tend to be a bit larger than females and may have wider heads and darker, more saturated colors. Males may spin in place with the tip of the tail nearly touching the snout. They hold a willing female in place by biting the nape of her neck and can be quite aggressive. The eggs are laid about a month after mating and may take three months to hatch. Few eggs reach full term, however, and little is understood about artificial incubation of this species. Typical eggs are about 1.5 inches long and an inch in diameter. Clutches laid outside in burrows dug by the female may have the greatest chance of producing young. Hatchlings are about 5 inches long and reach a foot in one year if they adjust to captive conditions.

PHOTO: I. FRANCAIS.

Portrait of an Egyptian Spiny-tail, *Uromastyx aegyptius*. The rounded snout is distinctive, as is the convex area at the middle of the upper lip.

THE FUTURE

Though large numbers of immature and adult Egyptian Spiny-tails currently are being imported, there is no guarantee that this will continue. There have been signs of late that Egypt is about to limit exports of much of its fauna, and it is quite possible that the trade in *U. aegyptius* will be limited. There also is the problem that these large lizards sometimes do not adapt well to captivity and often die before being given the proper housing conditions. Even if properly housed, many specimens never do adapt and die after a year or two. Insufficient numbers are being bred to supply the terrarium hobby.

INDIAN SPINY-TAILS

The Indian Spiny-tail, *Uromastyx hardwicki*, no longer is a common terrarium animal, but it probably is one of the best-known uromastyx in nature. In its rather wide range from southeastern Baluchistan, Pakistan, and neighboring Afghanistan eastward through a broad belt over northern central India, it may be abundant in large colonies in dry, rocky and gravelly (not sandy) prairies with sparse shrubs and dry grasses. There it lives in self-dug burrows that may extend from 2 or 3 feet to perhaps 10 or more feet below the surface, often with sharp turns and side passages to complicate their shape. It has been suggested that the Indian Spiny-tail is the ecological equivalent of the ground squirrels of the American Southwest, colonies serving to transform a tremendous number of leaves and seeds into subterranean fertilizer and seed beds. Colonies may consist of 50 or 100 animals of all ages in a relatively small area. There they tend to spend the hot hours of the day sunning in the mouths of their burrows and confidently searching for food. They are well-known to the locals as tame, tasty lizards that are easy to catch if you can run fast enough or can block a lizard's return to its burrow. The fat from the tails is a popular item of local cuisine and also is noted as an aphrodisiac. Needless to say, large numbers of Indian Spiny-tails cannot continue to exist anywhere near heavily settled regions.

DESCRIPTION

Uromastyx hardwicki is a large spiny lizard with a body length in adult males of about 7 to 10 inches and a tail about 5 to 8 inches long, giving a maximum size of about 18 inches. Females are only a bit smaller than males and just as heavily built. Like most other uromastyx, the body is somewhat flattened, the head is short, the tympanum is distinct with weakly denticulated scales at its front edge, and the sides of the neck and trunk have loose folds of skin without enlarged spiny tubercles. The thighs have large spiny tubercles, however. The tail is distinctive in being moderately oval in cross-section, though strongly depressed at the base, and evenly tapering to a pointed tip. It is covered above with about 34 to 36 rows of enlarged, spiny scales, each row separated by several (two to four, occasionally six) irregular rows of small,

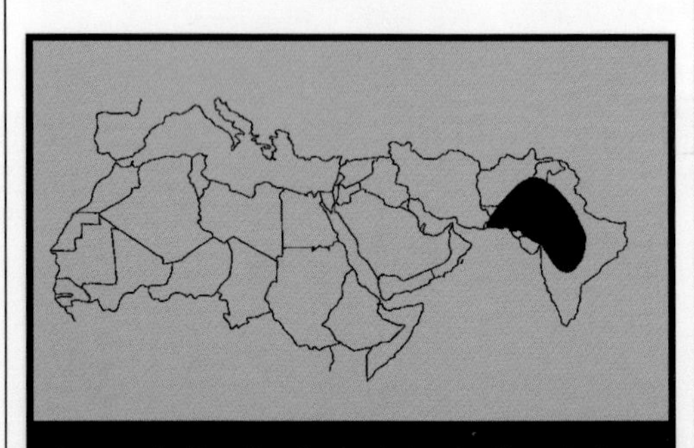

General distribution of the Indian Spiny-tail, *Uromastyx hardwicki.*

PHOTO: M. AND J. WALLS COURTESY ADAM'S PET SAFARI.

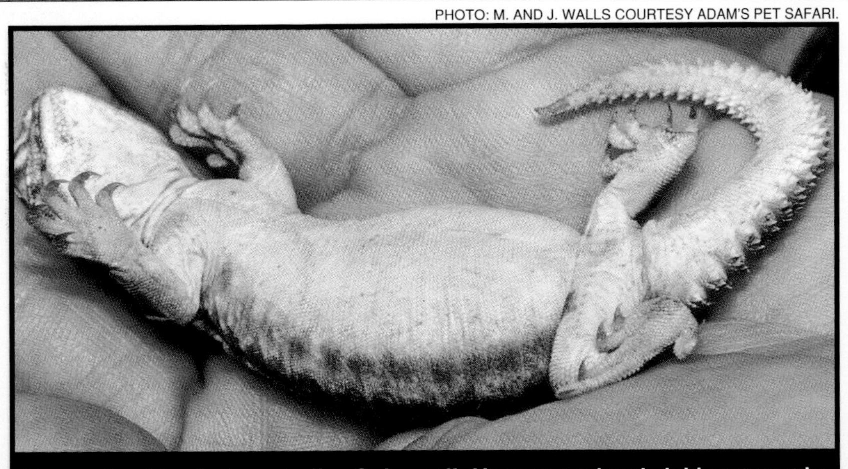

The underside of a baby Indian Spiny-tail, *Uromastyx hardwicki*, commonly is solid white. The large dark spot at the base of the thigh can just be seen.

normal scales (which may be hard to distinguish without magnification). The underside of the tail is covered with small scales much like those of the belly near the base. The absence of enlarged spiny tubercles on the body in combination with the presence of rows of small scales separating the spiny scales of the tail distinguishes *U. hardwicki* from the Asmussi Group of species found a bit to the west in similar habitat. It is believed that *U. hardwicki* is the most primitive species of the genus.

In color this is a rather plain species. It is sandy tan above, with a white throat and white belly. The back may be plain or marked with darker brown spots or a rather dense mesh of connected brown lines and mottling. There are no bright colors. One odd mark of the species is a large blue-black spot on the face of each thigh in the groin. Juveniles are darker brown with many blackish spots that are densest on the sides and may form one or two distinct dark stripes. They may have irregular white mottling on the sides of the head and shoulders.

There are about 15 to 18 femoral and pre-anal pores on each side of the body. There are no indications that the sexes can be distinguished externally by structure, so you must rely on behavior. Variation in this species

The tail spines of the Indian Spiny-tail are small compared to most others of the genus, are in more rows, and are separated by broad spaces containing several rows of small scales that may be barely visible in the young.

PHOTO: M. AND J. WALLS COURTESY ADAM'S PET SAFARI.

seems to consist mostly of differences in total length and development of large flat tubercles into regular rows across the back. No subspecies have been recognized.

BASIC CARE

Though Indian Spiny-tails generally are slow-moving lizards, they need a large terrarium. A couple of adults need a cage at least 4 feet long and proportionately high and wide. About 2 feet of the height should consist of substrate, preferably a loose sand and loam mixture. The surface can be covered with gravel, flat rocks, and sod, not loose sand. Provide the usual hiding places. Since 2 feet of substrate is not really enough for an adult to make into a comfortable burrow, it may be best to provide them with PVC pipes about 3 inches in diameter buried below the surface. The lizards will adjust to these quickly and build a chamber at the end near the bottom of the cage. The chamber should be relatively moist compared to the surface of the terrarium and should never be allowed to become too warm.

This baby Indian Spiny-tail, *Uromastyx hardwicki*, is about as pretty as the species ever gets. Though hardy, this can be a very difficult species to raise from wild-caught young.

PHOTO: M. AND J. WALLS COURTESY ADAM'S PET SAFARI.

Heating and lighting are much as for any other uromastyx. Provide full-spectrum fluorescent lights from mid-morning to late afternoon and a very hot basking area under an incandescent or metal halide lamp. Basking temperatures should be in excess of 95°F, with 120°F or a bit more quite satisfactory. One corner of the cage should never become much warmer than 85°F, however. At night allow the cage to drop to room temperature, as low as the mid-60's F. Short exposure to a UV light every week will help bring out colors and increase activity.

If at all possible, allow your lizards to bask outdoors in direct sunlight during the summer. They are fairly tolerant of short periods of high humidity but should not be left outdoors if the humidity is high for more than a couple of days. Remember that they are strong and fast burrowers and that they need to be able to retreat to a cool, shaded refuge.

In nature the Indian Spiny-tail becomes active about 10 or 11 AM and retreats into its burrow by 4 or 5 PM during the summer. It tends to brumate (limited hibernation) when outdoor temperatures drop much below 60°F, depending on fat stored in the tail to sustain it from November to February. Long-term health of the animals and breeding potential are improved if the lizards also are brumated during these months in captivity, but it is safer to provide a winter temperature of about 65°F during this period than try for a more natural lower temperature.

Adult Indian Spiny-tails are almost exclusively vegetarians. They prefer dry plants of various types (including brown rice and corn), but learn to take more moist vegetables in captivity. Some specimens like dandelion and carrot leaves, others like yellow and red flowers, and still others will feed on alfalfa hay. Few take moist fruits. Adults are excessively tame and seem to enjoy being handled (watch out for the spines on the thighs—they are harder to handle than those on the tail) and taking food from your fingers. There is no reason to feed adults any animal foods, and it might cause kidney damage. Juveniles, on the other hand, actively chase and eat grasshoppers and crickets as well as beetles.

BREEDING

There are few records of successful breeding followed by incubation of the eggs in captivity. In nature, the breeding season extends from perhaps late February or March into April, with the eggs laid from late April into June. The eggs are about 1 to 1.5 inches long, white, oval, and have thin shells. Clutches of 8 to 14 or perhaps 16 eggs are recorded. The eggs are laid in a chamber off the main burrow and develop there, probably under rather humid conditions. The young leave their mother's burrow from late June to July.

PHOTO: M. AND J. WALLS COURTESY ADAM'S PET SAFARI.

Indian Spiny-tails have very blunt snouts even visible in babies. Notice the large claws even in the young that are sharing burrows with their mothers.

Limited success has been attained by brumating healthy adults from November into February at about 65°F. The animals will become active occasionally during this period and should be allowed to come to the surface to sun themselves if they prefer. By early February or late January they may come up for an hour or two once or twice a week, and by the end of February they will be almost back to their normal activity cycle. If the adults are placed together then, males will begin a circling dance wherein they curve their body in nearly a circle and spin in place while marking their territory with a milky secretion from the cloaca. Females and even immatures also may circle, but the larger size and more determined behavior of the males should be obvious. If the female is willing to mate, the male nudges her with his snout and eventually bites her behind the head and twists his body under hers to insert a hemipenis into her cloaca.

Egg-laying occurs about 30 to 60 days later. The entire clutch appears to be laid in rapid succession. The female needs to be well-fed and rested after laying.

Incubation has been a stumbling block in successfully breeding these lizards. The eggs do not hatch if kept too dry, and they fungus if too wet. If you are lucky enough to get a clutch of eggs, your best bet may be to try incubation in moist vermiculite, trying different parts of water to base with portions of the clutch. Remember that the burrow is cooler and more moist than the surrounding habitat. The babies, if in good condition, can be described as cute and cuddly, but they are very shy and may not feed well.

PHOTO: A. NORMAN.

Adult Indian Spiny-tails, *Uromastyx hardwicki*, seldom are available today. They are the least spiny and most primitive of the *Uromastyx*.

THE FUTURE

Presently India and Pakistan do no export many reptiles to the American and European markets, and few specimens of *U. hardwicki* are available except to zoos and other institutions. The captive-breeding rate has been very low, so few captive-breds reach the hobby market. However, India may be loosening its export restrictions a bit, so we could see more Indian Spiny-tails on the market some day. Meanwhile, from an unexpected direction have come importations of hatchlings during July and August. These have come into the U.S. through the Ukraine, I am informed, and assumedly are being collected from western Pakistan or, more likely, Afghanistan (where the species occurs in the Kabul River valley). The babies I have seen have varied from near death to excellent condition, so a bit of careful judgment is required before purchase. Presently the species is under considerable pressure from local hunting for food and skins, but it lives in a large area that often is sparsely populated and hard to reach even with modern transportation.

THE BUTTERFLY AGAMAS

Though they commonly are imported, the beautiful butterfly agamas, *Leiolepis*, seldom make satisfactory pets. This largely is because they are kept under crowded conditions and provided with temperatures that are far too low to allow them to digest food. If kept in large, warm, terraria, especially outdoors, they are among the most colorful of the lizards, but they will never be easy pets.

RECOGNIZING THE BUTTERFLIES

The four or five species of *Leiolepis* are easy to recognize as a genus. They are rather large (10 to 20 inches total length), usually colorful, lizards with small, blunt heads, long tails, and a flattened body form. They lack all traces of crests on the body or nape, and the small, rather spiny scales of the tail tend to be in distinct whorls or rings. The body scales are granular and slightly keeled on the back, but the belly scales are distinctly larger. The scales of the head are small and irregular, and there is a large and distinct tympanum or ear opening that helps distinguish them from the earless western North American *Holbrookia* and allies that may be very similar in color pattern at first glance. All the species have pale spots (usually yellow to red) over the back and often three narrow yellow stripes as well. The sides of at least the males have black vertical bars often contrasted against reddish to bluish white. Females are less colorful than males, especially on the sides. Both sexes have femoral pores (more obvious in many males), but pre-anal pores are absent.

Like their close relatives the uromastyx or spiny-tails, butterfly agamas have specialized dentition. There are distinct incisors or cutting teeth (two above, one below on each side), canines or eyeteeth (one on each side in both jaws), and molars or grinding teeth (usually 12 above and 11 below on each side). If you were to compare the head of a butterfly agama with that of a small uromastyx, you could see the distinctive similarities between the two genera.

In *Leiolepis* the sides of the body have a loose skin fold that is connected to long, movable ribs. When the ribs are erected, the body becomes very disk-like (the colorful sides become the "wings of the butterfly" that provide the common name) and provides an extra bit of lift when the animals jump (as well as increasing the surface area of the body to allow rapid warming when basking in the sun). The hind legs are very long, heavily muscled, and powerful, and the lizards are accomplished runners and diggers both. Jumps of at least 3 to 6 feet have been seen by naturalists, and there are old, unconfirmed, and probably incorrect records of jumping glides over 30 feet long. The butterfly agamas occur in

PHOTO: M. AND J. WALLS.

This male Northern Butterfly Agama, *Leiolepis reevesi rubritaeniata*, displays the brightly colored skin of the posterior sides. Movable ribs allow the lizard to spread the skin and glide short distances as well as increase the basking efficiency.

relatively dry, often sandy, areas of Southeast Asia from eastern Indonesia to southern China. Though they often are common along the sea coasts, they also occur in dry mountains surrounded by rain forest.

GENERAL CARE

Virtually all the butterfly agamas available to the terrarium hobby are wild-caught, and there have been very few successful breedings in captivity. Like any other wild-caught adult or half-grown lizards, butterfly agamas should be carefully examined in the pet shop before purchase. Check for all the usual signs of problems, including watery eyes, raspy breathing, cottony growths in the mouth, poor muscle tone in the legs, obvious scars and external parasites, and thin, shrunken tail bases. Purchase the best specimens you can find and afford (currently butterfly agamas are relatively inexpensive lizards), and then take them to your vet to be checked and treated for intestinal worms.

THE TERRARIUM

In nature *Leiolepis* tend to be colonial, many animals occupying individual burrows within a small area of suitable soil type. They typically are non-aggressive, but each animal must have its own burrow into which to retreat and spend most of its hours. This lifestyle means that you can keep several specimens together in one terrarium, but it also means that you need to give them a special type of terrarium.

To do their best, give every two or three butterfly agamas a terrarium at least 3 feet long and preferably 6 feet. This means that you probably will have to build a terrarium to suit your specimens, though one or two juveniles can be kept for at least a while in a large all-glass terrarium. Metal cattle watering troughs have worked well for specimens kept outdoors in hot, dry climates. Additionally, the terrarium must be exceptionally deep, preferably at least 3 or 4 feet, to allow for a substrate of loose sandy soil at least 2 feet deep. Butterfly agamas each require a burrow about 2 feet long, often sharply angled before the end, and at least a foot under the surface. In this respect they are much like the uromastyx, so they can be given artificial burrows of PVC pipes.

The upper surface of the substrate should be dry, but the bottom must be fairly moist. A small pipe in each corner of the terrarium can be filled with water on a regular basis to control moisture at the bottom of the terrarium. You want it moist, so the animals can dig easily, but never muddy. Provide the usual hiding boxes, flat stones, and similar decorations, though the lizards will spend most of their time in the mouths of the burrows. Some hobbyists have recommended a covering of sod grasses as providing a very "homey" atmosphere for the lizards.

Unlike the uromastyx, butterfly agamas run and jump. If annoyed or surprised, they tend to run full speed into the sides of the terrarium or make a leap of several feet toward the rim of the cage. Obviously, glass terrarium sides will cause severe accidents and probably a few broken bones very quickly, so it is best to provide more flexible material such as soft mesh on at least two sides. Plants, rocks, branches, and similar items help give the lizards confidence. There must be a sturdy, well-fastened cage lid both to keep the lizards in and to hold the lighting system.

HEAT AND LIGHT

In these respects butterfly agamas are very much like spiny-tails. There should be a heating pad or heat tapes under the bottom of the terrarium to keep the burrows warm (about 75°F minimum during the day), and there must be a hot incandescent or even metal halide lamp at one end of the terrarium over a basking rock or rocks. The basking area should attain at least 100 to 110°F for several hours each day. It does no harm if the temperature locally reaches over 120°F occasionally. At night the temperature can be allowed to drop to the mid-70's F.

Provide the same type of lighting as for spiny-tails, preferably full-spectrum fluorescent reptile lights that extend over most of the terrarium. The air temperature should reach about 90°F during the day even away from the basking spot, but there must always be a bright yet cool corner into which the lizards can retreat.

OUTDOOR CAGING

Because of their high temperature, light, and space requirements, butterfly agamas are excellent lizards for an outdoor pen. As usual, the outdoor cage should be placed so it is in full sunlight most of the day yet has a shaded area. *Leiolepis* are not especially hurt by high humidity, which is a plus if you don't live in California or Arizona. In fact, they like a bit of condensation in the morning.

The outdoor pen should be at least 10 feet long, 4 feet wide, and 4 feet high for the lizards to be most comfortable. Because they will dig extensively, it would be best to dig out under the pen to a depth of 2 to 3 feet and line this area with mesh to prevent escapes. A more temporary but suitable setup would be to cover most of the bottom of the pen with rocks and provide a deep box of looser soil for burrowing. Either method should work.

If outdoor temperatures drop much below 85°F for more than a day or two, the lizards must be returned indoors to a heated terrarium or you must give them additional heating until the temperature rises again.

FOOD AND WATER

Though butterfly agamas are largely vegetarian, they usually will not turn down insects. In nature it seems that most feed largely on leaves, grasses, and fruit, and this should be their basic diet in the terrarium. Provide them with a good mixed veggie salad consisting of varied greens and fruits, and see what they prefer. Once a week supplement their salad with mealworms, crickets or grasshoppers, and waxworms. Juveniles should have a vitamin and mineral (calcium) supplement added occasionally. It shouldn't take you very long to determine the preferences of your pets. If they prefer meat to vegetables, it probably will do no harm to cater to their desires.

Because high humidity is not a great problem with these species, it does no harm to provide them with a small water bowl. They may or may not drink, as they get much of their water from their food. Specimens that feed mostly on insects should require more water than those that feed heavily on vegetables.

REPRODUCTION

The American whiptails (*Cnemidophorus*) of the Southwest are famed for the occurrence of parthenogenetic species, populations of apparently hybrid origin in which only females normally occur. In these parthenogenetics, males are absent or rare and perhaps sterile. Females may mate with males of other species or go through the motions of copulation with other females, activities that lead to the beginning of development of a fertile egg that eventually hatches into a female like her mother.

Something of the same situation occurs in at least some populations of butterfly agamas. This has greatly complicated their taxonomy and made the status of some of the described forms uncertain. Additionally, the tendency of these lizards to live in small, often isolated, colonies leads to many local variants.

As a rule, assuming both sexes are present in the species or population, males are a bit larger and usually much more colorful than females. They have bright patterns on the sides of the body and usually have bright body colors as well. Females may have the pattern of the sides reduced or missing, depending on species and subspecies. The bright side patterns are used by the males during courtship displays.

Virtually nothing seems to be recorded about the reproduction of the butterfly agamas in captivity, but it is known that in nature they lay only a few large eggs that hatch in the spring. A male and a female may form a bonded pair that share a burrow most of the time. The eggs are laid in a chamber off the burrow, and the young remain in the burrow with their mother until old enough to fend for themselves.

Because of the large accommodations needed to make these active lizards feel at home, plus the potential presence of parthenogenetic forms in the hobby, breeding butterfly lizards in captivity may never become commonplace.

THE SPECIES

Presently four distinctive species plus one questionable parthenogenetic form have been named; additionally, two of the species have described subspecies, so there is a total of seven named taxa in the genus *Leiolepis*. At least three species are present sporadically in the hobby.

Leiolepis belliana, the **Striped Butterfly Agama**, was (at least until recently) the most commonly seen species. In its typical subspecies, ***L. b. belliana***, from

General distribution of the Striped Butterfly Agama, *Leiolepis belliana*; *L. b. belliana* to the south, *L. b. ocellata* to the north.

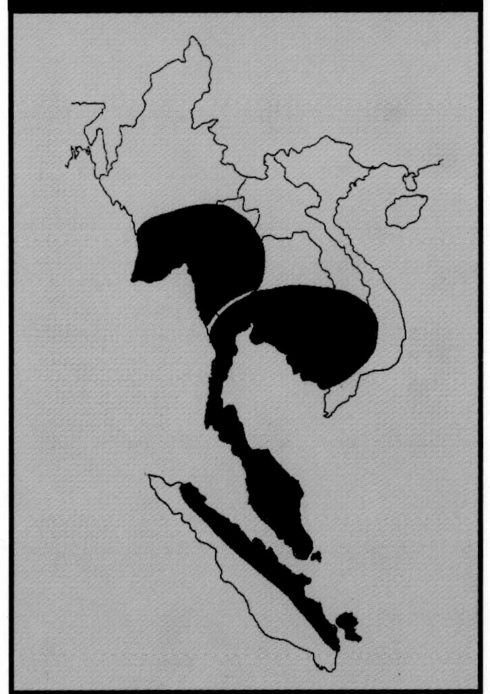

LEIOLEPIS Cuvier, 1829

Leiolepis belliana (Gray, 1827) .... Striped Butterfly Agama
Leiolepis guttata Cuvier, 1829.......... Giant Butterfly Agama
Leiolepis peguensis Peters, 1971 .... Pegu Butterfly Agama
Leiolepis reevesi (Gray, 1831) .... Northern Butterfly Agama
Leiolepis X triploida Peters, 1971 ..... Triploid Butterfly Agama

PHOTO: R. D. BARTLETT.

A male Striped Butterfly Agama of the northern subspecies, *Leiolepis belliana ocellata*.

the coastal region of Southeast Asia (from the Mekong delta through Thailand and Malaysia, south to Sumatra and Banka), this is a brightly striped and spotted lizard. Adults are brownish above with three narrow yellow stripes running from the rump to over the front legs, and there are two irregular rows of rather small yellow spots ringed with black on each side of the middorsal stripe. The sides of both sexes (but especially bright in males) are orange with some seven to nine distinct black vertical bars in high contrast. When specimens are sunning themselves with the body flattened, the colors of the sides are exposed to full view. The tail is reddish in juveniles. Additionally, in this species the sides of the tail may be brightly striped with yellow and rusty red in adults and there may be a few black bands on the upper front arms as well. *L. b. belliana* can be a truly stunning lizard when warm and given plenty of room.

Though most Striped Butterfly Agamas reproduce sexually, both sexes being present in a population, at least two instances are known of local populations that contain some parthenogenetic specimens. The name *L. X triploida* is based on such a local parthenogenetic form and may be the result of a successful mating between a normal male and a parthenogenetic female. Parthenogenetic forms may differ from sexual forms in coloration and size as well as details of scalation and chromosome counts. Much remains to be

The Striped Butterfly Agama, here a male of the southern subspecies, *Leiolepis belliana belliana*, currently is not being imported in large numbers.

PHOTO: G. DINGERKUS.

discovered about these attractive and often common lizards.

The subspecies ***L. b. ocellata*** Peters, 1971 comes from southwestern Burma, including the Pegu Island group, and northern Thailand, and tends to be found at much higher altitudes than the typical subspecies. In this form the yellow stripes are present only over the rump and the rest of the back is covered with large, black-outlined yellow spots in more or less distinct rows across the back. The spots may fuse into broken yellow cross-bands. Though the side pattern is similar to that of *L. b. belliana*, there really is no evidence that this is not a full species instead of a subspecies.

PHOTO: P. FREED.

A male of the southern subspecies of the Northern Butterfly Agama, *Leiolepis reevesi rubritaeniata*.

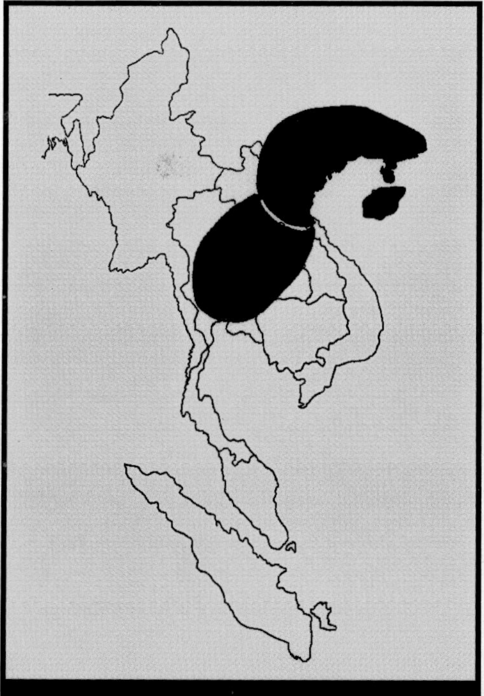

General distribution of the Northern Butterfly Agama, *Leiolepis reevesi*; *L. r. reevesi* to the north, *L. r. rubritaeniata* to the south.

Recently imports of the **Northern Butterfly Agama, *L. reevesi*,** have come in from Vietnam and southern China. This species lacks the stripes of *L. belliana* and sometimes has a different pattern on the sides. The sexes are very different in color, females lacking the contrasting black bars on the sides found in males and having overall duller coloration. The back is covered with round yellow spots that are very close-set and sometimes reduce the dark background color to a thin network. Males have orange sides that in the typical subspecies (***L. r. reevesi***) are covered with close-set black vertical bars. In the female the side pattern is absent or only weakly developed, but the upper side in front of the hind leg is reddish.

The poorly understood subspecies ***L. r. rubritaeniata*** Mertens, 1961 (from Laos and Thailand) has the pattern on the sides absent from the middle and present only behind the front legs and before the hind legs (only two to four black vertical bars in each position), and the pattern on the back mostly lacking yellow and presenting the appearance of a grayish reticulation or network. Females are more likely to have remnants of yellow spots on the back and have a broad reddish stripe on the upper side before the hind leg. There are few indications that this form actually is a subspecies of *L. reevesi*, and it may represent a full species.

The **Giant Butterfly Agama, *L. guttata***, recently has come into the hobby in numbers from Vietnam and perhaps northern Thailand. It is a heavy-bodied, very small-scaled species with a short head and long legs. The back is covered with many large and small pale spots on a bluish or grayish background rather than strong yellow ocelli surrounded with black. The sides of females are brownish, but in males they are barred in contrasting black and shiny white. Breeding males may have bright rosy tones over the head, legs, and upper sides. There may be traces of yellow stripes over the rump. The distribution and variation of this species, which adapts well to captivity, are poorly known. It is not especially longer than the Striped Butterfly Agama, incidentally, but heavier-bodied.

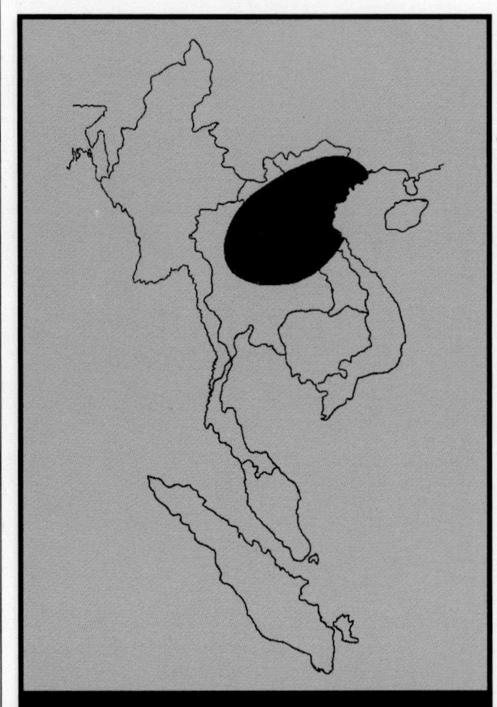

General distribution of the Giant Butterfly Agama, *Leiolepis guttata*.

PHOTO: P. FREED.

A stunning male Giant Butterfly Agama, *Leiolepis guttata*. Notice the black and white bars on the side without orange between them and the rosy tones on the head and neck.

The final species of butterfly agama, ***L. peguensis***, the **Pegu Butterfly Agama**, has a confusing

PHOTOS: M. AND J. WALLS.

In the female Northern Butterfly Agama, *Leiolepis reevesi rubritaeniata*, the pattern of the back is very subdued and there are no bars on the side, just an orange-tan blush. Males of course have a pair of blackish bars behind the front leg. This appears to be the common form imported at the moment into the United States.

array of characters and an uncertain distribution. Basically, it is patterned like *L. reevesi* above (many yellow spots ocellated with black and with little or no yellow striping) and an orange and black side pattern like *L. belliana*. However, the side pattern is reduced in extent, being mostly absent anteriorly, and the vertical bars often are short, irregular, and jumbled in arrangement. Unlike the other species, the femoral pores of males are tiny and often almost invisible except as fine dots rather than strongly marked pits. Described from the Pegu Islands, Burma, it probably also occurs in other parts of Southeast Asia. In many respects it seems to be a hybrid of *L. reevesi* and *L. belliana*, but there are supposed to be differences in scalation from those two species. Until the form can be studied in nature, it is best to retain it as a full species.

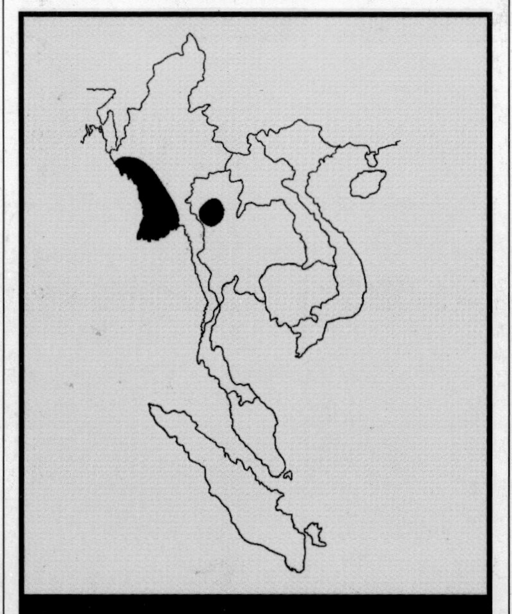

General distribution of the Pegu Butterfly Agama, *Leiolepis peguensis*.

THE FUTURE

As imports from southern China, Vietnam, and perhaps even Laos and Cambodia increase, it is likely that still more types of butterfly agamas will reach the hobby market, some of them possibly undescribed species. Hobbyists should keep careful notes and a photographic record of their purchases to allow definite identification at a later date if the taxonomy of this genus should be revised extensively in the future. When attempting breeding, be sure to try to get males and females of the same taxonomic form from a single importation to prevent possible genetic mismatching. Butterfly agamas are a very complicated group that are badly understood both by herpetologists and hobbyists.

Massive deforestation, land conversion, and a rapidly growing population throughout Southeast Asia, coupled with the residues of several recent wars, have destroyed much of the habitat that butterfly agamas need. Additionally, these lizards often are sold as snacks in the markets of Bangkok and other major cities of the area, and they are considered a tasty treat. Few imported specimens survive long in captivity because of stress and improper housing, so in addition to being taxonomically complicated these are tough terrarium charges.